WHERE IN
THE WORLD,
WHEN IN
THE WORLD?

WHERE IN THE WORLD, WHEN IN THE WORLD?

AN INTRODUCTION TO TRAVEL GEOGRAPHY AND INTERNATIONAL TIME

BEN GEORGE, Ph.D.

National Publishers
of the Black Hills, Inc.
Elmsford, NY

To Doug
A Fellow Traveler

Cover and Interior Design: Hudson River Studio
Typesetting: Techna Type, Inc.
Editing and Production: Joanne Modero Kelly,
Jane Andrassi, and Ellen Schneid Coleman
© 1988, National Publishers of the Black Hills, Inc.
All rights reserved.
ISBN: 0-935920-37-4
Printed in the United States of America.

CONTENTS

TRAVEL GEOGRAPHY

Where in the World?

What subject could be of greater importance or greater interest to travel industry personnel then geography? After all, underlying everything you do, especially as a travel agent—whether it is selling a cruise, writing an airline ticket, booking a hotel or organizing a tour—must be the knowledge of *where in the world* your client is going. And true professionals do not stop there. They know the location of places in relation to other places, the culture and climate of a particular country, and the major attractions of leading destinations.

As you can well imagine, the study of geography is a lifelong pursuit among true professionals, for no matter how much you know, there is always more to learn. Far from being a burden though, you will find that this quest is one of the aspects of the travel industry that makes the profession so interesting.

Where in the World, When in the World? has been designed as an introduction to world geography and international time for people in the travel industry, particularly for those, like you, who are just entering the field. It is, therefore, an introduction to travel geography with an emphasis on information of special interest to travel industry personnel. For more complete information on other areas of the world, consult David W. Howell's *Discovering Destinations.*[1]

In *Where in the World, When in the World?,* we shall provide you with a framework for continuous study of geography throughout your career. After introducing you to the travel industry and providing an overview of the world, we shall teach you the major physical features, regions, and countries of each of the inhabited continents. We will focus on an in-depth study of the United States and Canada as well as on some of the key tourist areas in the world. We also will discuss the locations of the world's best-known tourist attractions. In the last part of our text, you will learn the international system of keeping time called Greenwich Mean Time or GMT. As you continue to travel, to read, to attend seminars, and to learn in many ways, you will expand this framework with your own experience and knowledge.

Since many of you have not had any formal education in geography since junior high school, we'll start on a very basic level and move forward. If our starting level is too low for you, bear with us, use these first lessons as a review, and get ready to move forward when we reach your level.

Most importantly though, you should understand that *these lessons are designed to be studied with a good atlas in hand.* Using an atlas while you read will help you fully understand the material and will be needed to research many of the follow-up questions. As a travel professional, your atlas will be your best friend so it is appropriate that you begin learning to use it now.

Happy travels through *Where in the World, When in the World?*

---

[1]National Publishers of The Black Hills, Inc. 1987.

Photo: Courtesy Norwegian Caribbean Lines

THE TRAVEL INDUSTRY

An Introduction

Before embarking on our tour, however, let us take a brief look at tourism as an industry. As a professional, it is important that you understand the industry's worldwide economic significance, appreciate its rich and colorful history, and be aware of its bright expectations for the future. As you learn more about the industry and how it operates, you will see how vital information—geographic and other—is to the continued growth of the industry and your own success in it.

As a travel student you have taken the first step toward a career in one of the world's most interesting, important, and fastest growing industries. The travel industry provides career opportunities for millions of people and generates billions of dollars for salaries, taxes, and investments all around the world. Its continued growth has assured the prediction of many economic experts that travel would be the world's largest industry by the year 2000.

THE PRESENT

Today people travel everywhere from weekend getaways close to home to the far corners of the world. They travel for a variety of reasons and, consequently, travel has come to mean many different things to many different people. For some, travel is part of their jobs. For the salesperson, travel may mean "clinching an important account." For the teacher, travel may mean gathering information and inspiration for next year's classes. For the sales representative, travel may mean displaying a new line of merchandise at a trade show. For the professional, travel means networking with colleagues at a convention.

Many people, however, travel simply for pleasure and their reasons are as varied as people themselves. To the harried white collar worker, travel means an escape from stress, perhaps on board a luxurious cruise ship. To the bowling team or the square dance club, travel means competing in the national finals. For the newlyweds, travel means a honeymoon at a romantic resort. For the elderly, travel means Christmas with grandchildren. For the skier, travel means fresh powder on challenging slopes. For the shopper or collector, travel means entrance to the world's greatest emporiums. For the retired person, travel means a just reward for years of hard work. For the student, travel means seeing textbooks come alive. For the sports fan, travel means the thrill of being there. For the serviceman or woman, travel simply means going home.

Whether they are traveling to earn their livings, enrich their lives, or visit their families, the travelers of the world depend on the personnel of the travel industry at every phase of their journey, from planning and arranging a trip to providing the necessary services. As a professional in the travel industry, you will be helping these people meet their travel needs and get the most out of their trips.

Your work in the travel industry will also contribute to the vast economic importance of this industry. And as a travel student, it is important for you to have some idea of how significant the industry really is to the world's economy. Travel and tourism is already the leading industry in many nations of the world and is gaining rapidly in many others. The World Travel Organization (WTO) estimates that the 340 million people who traveled internationally in 1986 spent so many billions of dollars that travel has become the largest industry in the world. When international and domestic travel markets are combined, the expenditures of $2 trillion dollars[1] more than double those of world military spending. To see it in yet another way, world travel expenses are greater than the gross national product of any country in the world except the U.S.A.

In the United States alone, the U.S. Travel Data Center estimates that travelers spent some 492 billion dollars in 1986, making travel one of the leading industries of this country. In fact, according to the American Society of Travel Agents (ASTA), travel ranks as the first, second, or third industry in forty-six of the fifty states. States such as Florida and Hawaii depend heavily on the travel industry. Even in an industrialized state such as New Jersey, beautiful beaches draw so many vacationers that travel is the state's second largest industry.

The vastness and diversity of the travel industry mean jobs for many people. In the United States, one out of every fifteen people is employed in travel. As the industry continues to grow so do employment opportunities. Moreover, these jobs are widely distributed among all segments of the population. (In a recent year 1.2 million jobs were held by members of minority groups, four million by women, and 2.9 million by young people from ages sixteen through twenty-four.)

In addition to providing salaries for millions of Americans, the travel industry also paid more than twenty billion dollars in taxes to all branches of government in 1983. Understandably, then, the average state spends nearly one million dollars a year to promote tourism and the figures are growing. Let us take California as an example. The state spent $470,000 to promote tourism in 1984 but increased that figure to an astounding $5.7 million for 1985. Why? More than half a million Californians work in the travel industry and the industry brings in almost $900 million in state revenues a year. That means that the state earns back in three days what it spends in a year.

As important as travel and tourism is in both personal and economic terms, its significance extends far beyond to the most profound issue of our time—world peace. Many people have written of the value of travel in promoting international understanding, but President John F. Kennedy said it most eloquently:

> *Travel has become one of the great forces for peace and understanding in our times. As people move throughout the world and learn to know each other, to understand each other's customs, and to appreciate the qualities of individuals of each nation, we are building a level of international understanding which can sharply improve the atmosphere for world peace. . .The travel industry plays a key role in*

[1] Somers R. Waters, *Travel Industry World Year Book—THE BIG PICTURE* (New York: Child & Waters, Inc., 1987).

stimulating this flow of [the] world's peoples. . .a most important international activity. . .

As a future professional in the travel industry you may look forward to assisting people plan memorable travel experiences, participating in one of the world's strongest and fastest growing industries, and contributing to the international goodwill which may someday bring peace to all of us.

THE PAST

As a novice in this vast, important, and expanding industry, you may well be wondering, "How did it get to this point?" The answer to that question takes us down a long and colorful road with many detours, but we can safely simplify by saying that the growth and development of the travel industry closely follows technological developments which have made transportation faster, more comfortable, and less expensive. The roots of the travel industry may be traced all the way back to Biblical times and the empires of the ancient world. The modern travel industry, however, dates from the nineteenth century and the widespread use of the railroad and the steamship.

Both the railroad and the steamship extended the concept of travel for pleasure to the rising middle class of the day by providing comfortable means of travel and shortening travel time. Thomas Cook, considered the father of the modern travel industry, was quick to see the advantages of the new means of public transportation and booked his first package, an excursion to a temperance rally, on a railroad in 1841. In 1883, the *Orient Express,* destined to become the world's most famous train, had its inaugural run. In 1889, the *City of Paris,* a twin-propelled, steel-hulled liner, cut travel time between Europe and America to an astounding six days.

While the railroad and the steamship were altering the means of travel, Western European society was undergoing significant changes which would create the tourists who would use these new means of transportation. Starting in the late eighteenth century, the Industrial Revolution changed Europe from an agricultural world of villages and farms to an industrial world of cities and factories. This was to have a profound effect on travel and tourism.

This new world full of cities and factories soon created a need for escape from crowds and congestion, dull repetitive work, and industrial grime. It also created an urban middle class with leisure, discretionary income, and a desire to imitate the behavior of the aristocratic classes in many ways, including travel. Spas and beach resorts, where the waters were considered beneficial for health, were the first destinations. Hotels naturally followed. Here, then, are some of the roots of our present-day travel industry. Our urban life has increased in complexity to the point that many people now consider a temporary escape—a vacation—a necessity of life. Many people still look to the upper classes or, now, to the jet set or celebrities for guidance on travel plans. Beach resorts continue as top tourist attractions, emphasizing the benefits they have for our mental and physical health.

By the 1920s and 1930s, railroads, steamships, and hotels had reached such a dazzling level of elegance that those decades are considered the Golden Age of Travel. Soon, however, the world of the *Orient*

Express and ships of such great lines as the Cunard line were to give way to the automobile and the airplane. Together, they so revolutionized travel that they may be said to have "democratized" it. The automobile decreased the cost of travel, thereby making it available to more people; the airplane decreased the time needed to get from place to place, enabling the worker with limited time to take advantage of distant vacation spots.

Air travel had already made inroads into the travel industry by the 1930s. Aviation technology developed by the military in World War II helped commercial aviation replace railroads and steamships as the primary means of long distance transportation over land and sea. Despite the steamship's many appeals, it could not match the speed or the relatively low cost of travel offered by jets. In 1955, more Americans crossed the Atlantic by plane than by ship and the world watched the demise of the once mighty transatlantic liner industry. Many of the ships were later restored and pressed into service along with newer ships for the booming cruise market of the 1970s and 1980s—a brilliant example of industry resilience. On land, automobiles as well as airlines led to the demise of the passenger rail industry and the famous trains of earlier years.

Modern aviation now provides a twentieth century magic carpet to the travelers of the world, whisking them in speed and comfort to distant corners and opening places only dreamed of before. Besides providing transportation for millions of people, the airplane, together with the automobile, has also stimulated the growth and development of other branches of the travel industry including hotels, motels, resorts, theme parks, commercial recreation sites, and sightseeing and tour companies. The number of travel agencies has also grown in response to the growth of the industry. At the end of World War II, there were 1,000 agencies in the United States. Today there are almost 30,000.

While developments in transportation were essential to the growth of the travel industry, social and cultural changes also played a major role. For example, the paid vacations which labor unions established have enabled working people to enjoy the pleasures of travel. Medical science has prolonged our life spans and so today the United States has more retired people than ever, and retirees are now among the most frequent travelers. New waves of immigrants maintain family ties with relatives all over the world. The women's liberation movement has freed women to travel comfortably alone. Recreation and leisure are now accepted as necessities of American life. All of these social and cultural changes and many more have expanded the market for travel industry products.

Technological advances continue to affect the travel industry. Widespread air conditioning has extended the season in many resort areas. Television also stimulates people to travel by broadcasting specials on specific areas of the world, by using exotic, foreign backgrounds in such shows as *Magnum, P.I.,* and by promoting shows based on the world of travel such as *Love Boat*. The future cannot, of course, be separated from the world of space travel and we already have at least one company taking reservations for the first tourist excursions into outer space.

THE FUTURE

As a student in the travel industry, the future is yours. All signs indicate that the travel industry will continue to grow well into the future. Continued growth means, of course, many employment opportunities for those just entering the industry. Given the size and complexity of the industry, travel and tourism can offer a very wide spectrum of positions utilizing a variety of skills, knowledge, and talent. These career opportunities have been well described in *The World of Travel.*[2]

Since the business of the travel industry is travel, many positions involve some business travel as part of the job. Even if travel is not a part of a particular job, employees often are eligible for travel discounts which make travel available at a fraction of the normal cost. Sometimes, such benefits are also available to members of the employee's immediate family.

Most entry-level jobs offer salaries comparable to those of other industries, in addition to the unique fringe benefit of discounted travel opportunities. Employees are not tied to entry-level jobs nor are they "locked in" with any given employer. People in the travel industry find that promotions within a company and/or job opportunities in other companies are readily available.

Below is a partial list of career opportunities within the travel industry. Some of these jobs require on-the-job experience and formal training. In addition to the positions listed below, employment is available in many areas of city, state, and national governments. Many foreign countries maintain offices in major cities to inform tourists of travel opportunities in their countries. State travel and tourism departments and regional travel associations promote travel to various destinations in the United States and the world. The list below is not a total or complete list of all career opportunities; however, it provides an indication of the variety of positions available.

ENTRY-LEVEL JOBS: TRAVEL AGENCIES
Travel Counselor
Office Supervisor
Assistant Controller
Advertising Manager

CAREER JOBS: TRAVEL AGENCIES
Branch Agency Manager
Agency Owner
Assistant Manager
Marketing Manager
Controller
Outside Sales Representative
Supervisor, Sales & Service
Manager—
    International Department
    Domestic Department
    Tour and Group Sales
    Commercial Department

TOUR COMPANY JOBS
Manager
Tour Director
Purchasing Agent
Group & Charter Sales
Sales Representative
Controller
Assistant Manager
Tour Planner
Sales Director or Manager
Assistant Sales Director
Advertising Manager
Office Supervisor
Personnel Officer

AIRLINES
Flight Services
Customer Service
Operations
Dispatcher
Dispatch Clerk
Transportation Agent
Flight Attendant

[2]National Publishers of the Black Hills, Inc., 1979.

CARRIERS

City or Area Manager & Sales
   Director
Director, Passenger Sales
Advertising Manager
Zone Sales Manager
Reservationist
Interline Sales Representative
Director, Agency Relations
General Manager Agency Sales
Director, Group & Charter Sales
Operating Manager
Purchasing Agent/Controller

TRAVEL ORGANIZATIONS

Membership Secretary
Sales Representative
Research and Analysis
Regional Sales Office Manager
Convention and Business Travel
   Director

Director, Travel Information
   Services
Marketing Manager

**TRAVEL RESEARCH
AND DEVELOPMENT**

Statistical Analyst
Tourism Specialist
Management Consultant
Development Planner
Research Analyst

AIRLINE SECURITY

Passenger Security
Ground Hostess
Passenger Service Representative
Gate Agent
Transportation
Travel Clerk
Travel Agent Counselor
Ticket Sales Agent
Ticketing Clerk
Reservationist

Other travel industries that offer employment opportunities include: hotels, resorts, car rental companies, travel associations, travel publications, and tourist departments of city Chambers of Commerce.

Below are brief descriptions of some of the more popular entry-level positions in the travel industry.

Reservationist. A reservationist is usually the first person in the travel industry to come into contact with customers. Reservationists assist passengers by answering questions and booking travel and accommodation reservations.

Travel Counselor. A travel counselor works in a travel agency and arranges "package" vacation plans, cruises, tours, and business trips. Many of these tours involve groups of ten, twenty, or more passengers. The group is usually accompanied by a tour guide, who travels with the group to its destination country and handles all of the travel arrangements for the group.

Ticket Agent. A ticket agent works directly with customers at airports and city ticket offices of airports, railroad, and bus carriers. Responsibilities include writing tickets and assisting passengers in boarding. The ticket agent meets an interesting cross section of people from countries throughout the world.

Passenger Service Agent. A passenger service agent is responsible for taking care of the special needs of passengers who have never traveled before or who have problems created by handicaps, small children, etc. The passenger service agent also pays special attention to very young or very old people who travel alone as well as to celebrities and other important travelers.

Operations Agent. This position involves several different jobs, all of which relate to working in close proximity to aircraft. Examples are ramp agent, station agent, and cargo agent.

Flight Attendant. This person works for one of the national or international airlines and is involved in serving the needs of in-flight passengers. The flight attendant, like many other travel employees, wears a uniform specified by the employer and receives special orientation training in flight attendant duties. Each individual airline has a training program to train the flight attendant about its rules, regulations, and passenger service.

The flight attendant is an aircraft crew member who is assigned to duty in the cabin of an airplane to serve the comfort and convenience of passengers and maintain the security of the cabin during flight.

Hotel and Resort Employees. The variety of positions available in hotels and resorts is almost endless. Large hotels have reservationists and many other entry-level jobs as well as managers for each specific department within the hotel or resort.

Regardless of which of these careers you decide to pursue in the travel industry, you will need strong training in the basics of the industry. And nothing is more basic to travel and tourism than a solid knowledge of geography.

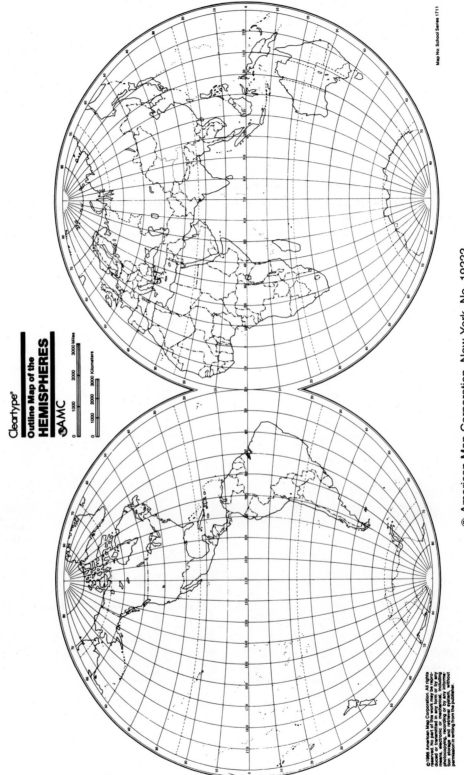

Clearlype®
Outline Map of the
HEMISPHERES
AMC

1000   2000   3000 Miles

1000   2000   3000 Kilometers

Map No: School Series 1711

© American Map Corporation, New York, No. 19222

AN OVERVIEW OF THE WORLD

Look at any map or globe and you will see that the surface of the earth is covered with large masses of land and water. The large landmasses are called *continents.* There are seven of them: *North America, South America, Europe, Asia, Africa, Oceania,* and *Antarctica.* Since Antarctica is uninhabited except for a few scientific stations, we shall exclude it from any further discussion.

Each continent is composed of offshore islands as well as its great landmass. Japan, therefore, is a part of Asia; the British Isles, a part of Europe. Every island in the world, then, is assigned to one of the continents. Indeed, Oceania is made up entirely of islands, including Australia which is sometimes considered a continent in itself.

The continents are grouped into *hemispheres,* literally meaning half of the world. North America and South America, joined by a thin strip of land called the *Isthmus of Panama,* make up the *Western Hemisphere.* The interconnecting landmass of Europe, Asia, and Africa along with the islands that make up Oceania are all in the *Eastern Hemisphere.*

The earth is also divided into *northern* and *southern hemispheres* by the *equator,* which is the imaginary line drawn around the earth at its widest point (24,900 miles). Logically enough, the area of the world north of the equator is the *Northern Hemisphere;* the area south of the equator is the *Southern Hemisphere.* Seasons in the Southern Hemisphere are exactly the opposite of those in the Northern Hemisphere with winter occurring in June, July, and August; spring in September, October, November, and so on. North America, Europe, and the landmass of Asia (except some of its islands) are in the Northern Hemisphere. The equator cuts through both South America and Africa leaving them divided between the Northern Hemisphere and the Southern Hemisphere. All the islands of Oceania are in the Southern Hemisphere.

All the land of the world, including the continents and islands, covers about thirty percent of the earth's surface. The other seventy percent of the earth's surface is covered by water. The large bodies of water are all called *oceans.* There are four: the *Pacific,* by far the largest, which contains forty-six percent of the earth's water; the *Atlantic,* the *Indian,* and the *Arctic.* There are many other smaller bodies of water which may or may not be connected to the oceans. These are bays, gulfs, seas, and lakes such as Hudson Bay, the Gulf of Mexico, the Mediterranean Sea, and Lake Superior. Think of them as you would extensions of the landmasses into peninsulas, isthmuses, or islands.

The earth is inhabited by about five billion people. About sixty percent of them live in Asia, and another fifteen percent live in the adjoining continent of Europe. Three-quarters of the world's people, then, live in the Eurasian landmass. And when Africa is included, the percentage of people living in the Eastern Hemisphere goes up to over eighty-five percent. In contrast, only fourteen percent live in the entire Western Hemisphere. Oceania, with less than one percent of the world's population, hardly alters the statistics.

The people of the world have organized themselves into 170 independent nations and about 40 territories not generally recognized as independent. The largest by far in area is the Soviet Union which spreads across all of Northern Asia and a large part of Europe, encompassing over eight and one-half million square miles. It is more than twice as large as Canada, which is next in size.

China, the United States, and Brazil rank third, fourth, and fifth respectively. Australia, India, and Argentina, in that order, round out the list of countries with over one million square miles. Five of these giant countries also rank very high in population. China has more people than any other country, followed by India. The Soviet Union ranks third; the United States, fourth, and Brazil seventh among the world's most populous nations. Indonesia and Japan, not on our list of the largest countries in landmass, rank fifth and sixth respectively in population.

The wealthy of almost all countries travel, but most of the world's travelers and tourists come from the developed areas of the world: (1) North America (the United States and Canada), (2) Western Europe, and (3) Japan. Although fewer in number, the citizens of Israel, Australia, New Zealand, and South Africa also travel extensively. There is also a growing number of travelers from the rapidly developing countries of Asia such as Taiwan and South Korea.

Because of its large population and its wealth, the United States commands a huge portion of the world's travel business, as airline statistics demonstrate. Consider that about forty percent of the world's air travel is United States domestic travel (within the United States and the lower 200 miles of Canada). About half of the remaining sixty percent of international travel involves flights coming from or going to the United States. The last thirty percent involves international travel to and from areas outside the United States. Understandably, then, a list of the ten busiest airports in the world shows that most are located in the United States with Chicago's O'Hare leading the way. The two foreign airports on the list are in London (Heathrow) and Tokyo, the gateways to Western Europe and Japan, the other two great sources of travelers and tourists in the world.

The world's travelers and tourists primarily visit places in their own countries or in other developed countries for business, family gatherings, pleasure, health, and education. They also travel to key commercial centers such as Hong Kong and Singapore, to selected sun spots such as Mexico and Tahiti, to points of cultural or religious interest such as the antiquities of Egypt or the city of Bethlehem, to points of natural or recreational interest such as the wildlife reserves of Kenya or the Great Barrier Reef of Australia, and to attend events such as Carnival in Rio, the Olympics, or a convention. Regardless of where they travel or why, the travelers and tourists of the world have made the travel industry one of the largest and fastest growing industries in the world.

In travel geography, as we shall see, it is not the size of the population or its area which marks a country's importance. Rather, it is the extent of the country's attractions—scenic, cultural, commercial, etc.—which make it important to the world's travelers and tourists. Those countries which are the favored destinations receive the most attention in travel geography. For this reason, we shall focus our attention on those parts of the world that are of primary importance to the travel industry. And there is no better place to start than with our own home continent, North America.

The Trans-Canada Highway—Five thousand miles from Atlantic to Pacific
Ocean. *Photo: Courtesy Canadian Government Travel Bureau*

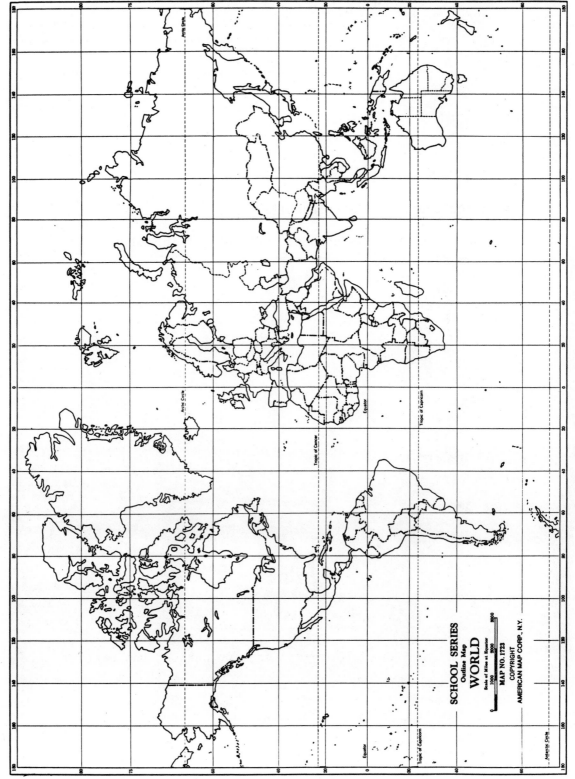

SCHOOL SERIES
Outline Map
WORLD
Scale of Miles at Equator
0   1000   2000   3000
MAP NO. 1723
COPYRIGHT
AMERICAN MAP CORP., N.Y.

AN OVERVIEW OF THE WORLD

1. What percent of the earth's surface is covered by land?

   _____

2. What percent of the earth's surface is covered by water?

   _____

3. How many continents are there? _____

4. Which continent is uninhabited? _____

5. Which continent is composed entirely of islands?

   _____

6. Name the two continents in the Western Hemisphere.

   _____

   _____

7. Name the land bridge that connects the two continents
   of the Western Hemisphere. _____

8. Name the continents in the Eastern Hemisphere.

   _____

   _____

9. How long is the equator? _____

10. Name the four oceans of the world. _____

    _____

11. Which ocean is the largest? What percent of the earth's
    water does it contain? _____

12. What is the world's population? _____

13. Which continent has the largest population? What per-
    cent of the world's people live there?

    _____

14. Which inhabited continent has the smallest population? What percent of the world's people live there?

_____

15. What percent of the world's people live in the Western Hemisphere? _____

16. How many nations and territories were there in the world as of 1984? _____

17. Which country is the largest in area? _____

18. Where does the U.S. rank in size of area? _____

19. How many countries have an area of over 1 million square miles? _____

20. What is the most populous country in the world?

_____

21. Where does the U.S. rank in population? _____

22. Where does Brazil rank in size? _____ In population?

_____

23. From which three areas of the world do most of the world's travelers and tourists come?

_____

24. What percentage of the world's air traffic is domestic United States travel? _____

25. What percentage of the world's air traffic is international? _____

26. Name the world's busiest airport. _____

27. Name the two foreign airports on the list of the ten busiest in the world. _____

28. What is the basis of a country's importance in travel geography? _____

_____

_____

29. Give as many reasons as you can for why people travel.

_____

_____

_____

_____

_____

30. Name the one place in this world where you would most like to travel and explain why. _____

_____

_____

_____

_____

NOTES

CHAPTER

1

NORTH AMERICA

CANADA AND ALASKA

As our home, North America seems to be the most familiar continent. Yet, it stretches to almost unknown corners of the world and has an amazing diversity in natural regions and cultures. On its northern edge, North America breaks up into the frozen islands of the *Arctic Ocean* and its long arm, *Hudson Bay.* These islands, sparsely inhabited by Eskimos, if inhabited at all, neighbor the giant landmass of *Greenland* (also part of the North American continent). The entire western coast of North America is bordered by the *Pacific Ocean,* the eastern coast by the *Atlantic Ocean* and the southern coast by the *Gulf of Mexico* and the *Caribbean Sea.* In the south, this great continental landmass narrows into a thin land bridge, the *Isthmus of Panama,* which connects it to *South America.*

The islands of the Caribbean Sea, the *West Indies,* with their sunny climate, lush foliage, and colorful Afro-European cultures, offer a striking contrast to the islands of the Arctic Ocean. In addition to these two groups of islands and Greenland, North America also includes the *Bahamas* and *Bermuda* on its Atlantic side.

The giant countries of *Canada* and the *United States* occupy much of the territory of North America. They are usually grouped together and distinguished from the other countries of North America as *Anglo-America* for several reasons: (1) they are neighbors; (2) they share a British heritage; (3) they have achieved many technological advances; (4) they have a high standard of living.

The other countries of North America, from Mexico south and including the West Indies, share, for the most part, a Latin (Spanish, Portuguese, French) heritage and are often grouped with the countries of South America. *Mexico* is the largest and most important of the Latin countries of North America. Between Mexico and South America, there are seven small countries which are collectively called *Central America. Mexico,* the *West Indies,* and *Central America* together with *South America* make up *Latin America,* a cultural term referring to all the Latin countries of North and South America.

As our "home" continent, North America is the focus of all of our domestic travel. Furthermore, the sun spots of Mexico and the West Indies draw so many vacationers that much international travel also occurs within North America. Because of the special importance of North America, we will treat Canada, the United States, Mexico, and the West Indies as separate units rather than studying the continent as a whole.

CANADA

Canada is one of the largest countries in the world, second in size only to the Soviet Union. Although the country is vast, much of its land is uninhabitable, and so its population is quite small (about one-tenth that of the United States) in relation to its size. Moreover, most of Canada's people live in cities spread from coast to coast along its southern border, making Canada a "horizontal" country with most traffic moving along an east-west axis. All but one of her cities, Edmonton, Alberta, are within two hundred miles of the American border making eighty percent of Canadians close neighbors of the United States.

While the two countries have much in common, Canada is distinguished from the United States by its close ties to Britain (it is a Commonwealth country), its distinctive French heritage which has created a unique subculture, its relatively homogeneous population (especially in comparison to the United States), its low population density, its vast tracts of wilderness, its severe climate, and its unique physical features.

The most distinctive and the most important physical feature in Canada is the *Canadian,* sometimes called the *Laurentian, Shield.* A remnant of the ice age, the harsh land of the Canadian Shield surrounds Hudson Bay like a giant horseshoe and extends as far south as the Great Lakes, making up about one-half Canada's territory.

Some parts of the Shield are barren rock; in others, there are pine forests and lakes. Although the Shield is rich in minerals, its climate is severe and its land is unsuitable for agriculture, and so it has never drawn much human settlement. Today, it is a vast empty place which effectively divides settled Canada into eastern and western sections.

The settled areas of Canada are usually divided into six regions: the Atlantic Provinces,[1] Quebec, Ontario, the Prairie Provinces, the Pacific Region, and the North. The Atlantic Provinces, Quebec, and Ontario are located to the east of the Canadian Shield. The Prairie Provinces and the Pacific Region are to the west. The North is usually identified as that sparsely inhabited area north of both the Prairie Provinces and the Pacific Region.

In the east, the mighty *Saint Lawrence River* flows thousands of miles from the Great Lakes to the Atlantic, providing a great natural east-west highway. This river has long been the focal point of settlement and development in Canada and it provides a unifying link between Ontario, Quebec, and the Atlantic Provinces.

ONTARIO

Let us start at the headwaters of the St. Lawrence with the province of *Ontario.* The very heartland of English-speaking Canada, Ontario is the wealthiest and most populous of all the Canadian provinces. Centrally located, Ontario contains much of the Canadian Shield north of the Great Lakes as well as the richly developed *Niagara Peninsula,* that bridge of land between Lakes Huron, Erie, and Ontario.

English-speaking *Toronto,* the major city of Ontario, is Canada's main air transportation center and its principal city in finance, commerce, communications, and industry. It is a lively city full of many colorful, ethnic neighborhoods, restaurants, shops, and night life. Toronto attracts many tourists as well as business travelers. While well known for its modern city hall and shopping galleria called *Eaton Centre,* Toronto's most visible structure is the *CN (Canadian National Railroad) Tower.* At 1,815 feet, twice the height of the Eiffel Tower, it is the tallest free-standing structure in the world. It contains shopping and dining as well as viewing areas.

The city's other well-known attraction, *Ontario Place,* built on stilts above Lake Ontario, draws many people to its ninety-six acres of parks, restaurants, theaters, and marina. As the location of Canada's busiest

---

[1] Canada uses the word province rather than state.

airport, Toronto is also one of the key transportation centers in North America.

Ontario also contains the national capital, *Ottawa.* Most visitors to Ottawa enjoy touring the old world parliament building, excellent museums, springtime tulip gardens, and witnessing the pageantry of the changing of the guards. Popular *Niagara Falls,* on Ontario's border with New York State, is best seen from the Ontario side. In the interior of the province, hunting and fishing on remote lakes draw many sportsmen.

QUEBEC

Moving down the St. Lawrence, we come to the province of *Quebec.* Quebec's French heritage distinguishes it from the rest of Canada. That heritage makes Quebec unique in North America for it has survived and flourished as an island of French culture in the middle of Anglo-North America.

Nowhere is the visitor more aware of the strength of the French heritage than in *Quebec City.* This historic city, the only walled city in North America, seems to exist in a vacuum of time and space and is truly an island of the Old World in the New. *Hotel Frontenac,* one of the castle-like hotels the Canadian Pacific Railroad built earlier this century, towers above

Quebec City, Quebec

the old city and adds much to its ambience. Winter comes alive in this city through its festive *Winter Carnival.*

Montreal is not only the largest city of Quebec, it is also the second largest French-speaking city in the world, and it delights visitors with its mix of modern urban amenities and Old World charm. Amid the skyscrapers, major league sports, and cultural complexes, Montreal has preserved its historic old town and built a seven-mile underground city along its subway route and filled it with shops, cinemas, bars, and restaurants. Visitors may take Le Metro, as the subway is called, to *Man and His World,* a theme park which grew out of Canada's centennial celebration, *Expo,* in 1967.

In addition to Quebec City and Montreal, Quebec also offers ski holidays in the *Laurentian Mountains,* cruises on the St. Lawrence River, stunning autumn foliage, the picturesque scenery of the *Gaspé Peninsula,* a number of religious shrines, and quaint Old World villages.

THE ATLANTIC PROVINCES

The islands and peninsulas surrounding the *Gulf of the St. Lawrence* are occupied by four provinces: *Newfoundland* and its mainland annex *Labrador, Nova Scotia, Prince Edward Island,* and *New Brunswick.* These are known as the *Atlantic Provinces.* Three of the provinces, Nova Scotia, Prince Edward Island, and New Brunswick, are sometimes called the *Maritime Provinces.* All four provinces have much in common with their neighbors, the New England States, and those features are the basis of the region's appeal: sites rich in colonial history, quaint and charming villages such as the much photographed Peggy's Cove, Nova Scotia, picturesque scenery, rugged seacoasts and beaches, sumptuous seafood, and seaside resorts. *Halifax, Nova Scotia* is the major city of the region and *Fort Louisburg, Nova Scotia,* a recreation of a colonial fort, is an especially worthwhile attraction.

THE PRAIRIE PROVINCES

Now turn to the vast prairies of Canada that spread west of the Canadian Shield to the Rocky Mountains. This area is one of the great breadbaskets of the world. Three provinces—*Manitoba, Saskatchewan* and *Alberta*—occupy the prairies and are, appropriately enough, called the *Prairie Provinces.* They have much in common with the Midwest of the United States, including vast grain farms and small farm towns.

Of the three, *Alberta* attracts the most visitors. Alberta is really more than a prairie province with a Midwestern life style. Located at the western end of the prairies, where the grasslands rise up to the great *Rocky Mountains,* it is Canada's wild west. For the visitor, this is best exemplified by *Calgary,* which celebrates its cowboy heritage with a giant rodeo, the *Calgary Stampede,* each summer.

Calgary is the foremost city of the Prairie Provinces. It is the center of the oil and gas industry that has made Alberta so wealthy and draws many business travelers. It is also the gateway to the parks of the Canadian Rockies.

The two best-known national parks, *Banff* and *Jasper,* protect much of the Canadian Rockies. Both offer spectacular scenery, hiking, camping, and trail rides in the summer and superb skiing in the winter. The *Banff-Jasper Highway,* winding through some of the world's most spectacular mountain scenery, links the two parks. The *Columbia Glacier,* probably the world's most accessible glacier, comes right down to the highway, allowing many visitors to enjoy an exciting snowmobile ride across it.

The village of *Banff* is the center of activity in these parks. Its most famous hotel is the *Banff Springs Hotel,* and it is also within easy reach of Lake Louise and the château on that lake. Both the Banff Springs Hotel and the *Château Lake Louise,* like the Hotel Frontenac in Quebec City, were constructed by the Canadian Pacific Railroad earlier in the century.

PACIFIC REGION

Mountains stretch from north to south all the way from Alberta to the Pacific Ocean making *British Columbia,* the only province in the *Pacific Region,* Canada's most mountainous province. Given the mountainous terrain of its interior, most of British Columbia's population lives in the metropolitan *Vancouver* area or on the offshore island also called, with some confusion, *Vancouver Island.* Vancouver Island is the largest island off the west coast of North America.

The city of Vancouver, on the mainland of British Columbia, is Canada's third largest city and its major west coast port and gateway. Tourists here enjoy the mild, but wet, climate of the Pacific Northwest. Vancouver offers one of the most beautiful settings of any city in the world, for it is wedged in a coastal plain between snow-covered mountains and sea.

Vancouver offers all the modern, urban amenities, a restored older section called *Gastown* and many parks and gardens. *Stanley Park,* the most famous, juts out from the city into the harbor. Nearby *Whistler Moun-*

Vancouver, British Columbia

tain provides fine skiing close to the city. Vancouver is important to the travel industry not only because it is a major destination in itself and a gateway city to the Orient, but also because it is one of the major departure ports for cruises to Alaska.

Vancouver Island's main attraction is the city of *Victoria,* capital of the province of British Columbia. Victoria brings a little bit of Old England here to the New World, and is known for its stately parliament building, a fine provincial museum, and the *Empress Hotel,* another of the great Canadian Pacific hotels. Traditionally, visitors to Victoria are fond of taking tea at the Empress and touring its gardens, especially the world-famous *Butchart Gardens* just outside the city.

THE NORTH

The last region of Canada, the *North,* composed of two territories, the *Northwest Territories* and the *Yukon Territory,* appeals to hardy sportsmen and adventure travelers. The Yukon is also visited by many tourists in conjunction with trips to Alaska.

ALASKA

Alaska, America's great wilderness state, draws tourists to its spectacular scenery and "last frontier" ambience. Many tourists visit Alaska by cruise

Alaskan Glaciers. *Photo: Courtesy Bradley ● McAfee Public Relations, Alaska DOT photo by Mark Skok*

ships through the *Inside Passage,* one of the world's most beautiful and popular cruise routes. The Inside Passage is the name given to the waterway between the mainland of British Columbia and Alaska and their many offshore islands. The islands protect the waterway from the full thrust of the ocean, making for smooth sailing. Day after day, passengers are treated to snow-covered mountains, deep green forests, glaciers, eagles, seals, and other wildlife. Historic towns such as *Sitka,* the former capital of Russian America, and *Skagway,* a jumping off point for those eager to participate in the Yukon gold rush, are popular sights.

The highlight of the cruise, though, is *Glacier Bay,* a National Monument where many glaciers flow into the sea in one of the world's most beautiful natural settings. *Mt. McKinley* or Danali, as the locals call it, is the highest mountain in North America and a major tourist attraction. A vast national park surrounds the mountain. There, nature enthusiasts can see a wide range of North American wildlife including mountain sheep, grizzly bears, caribou, and wolves in their natural habitat. Many visitors to the park arrive by a colorful railroad which connects the state's largest city, *Anchorage,* on the coast, with *Fairbanks,* the major city in the interior. Given its vast wilderness, Alaska naturally draws many sportsmen, campers, and adventure travelers.

Having arrived in the United States now, we'll continue touring the country through the "lower 48," as Alaskans call them, in our next chapter.

Totem Poles, Stanley Park, Vancouver, British Columbia. *Photo: Courtesy Government of Canada, Regional Industrial Expansion*

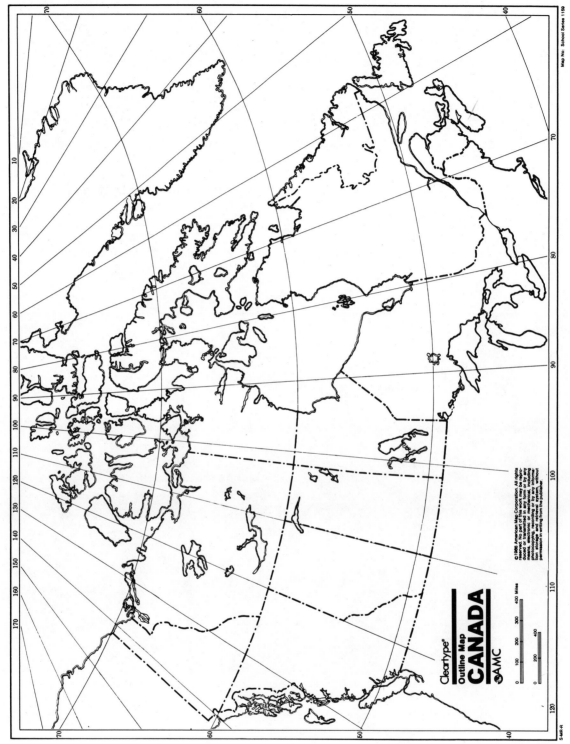

Cleartype®
Outline Map
CANADA
⊗AMC

Map No: School Series 1159

© American Map Corporation, New York, No. 19222

S-44R-R

NORTH AMERICA
Canada and Alaska

1. Name the three major countries of North America.

    _____

2. Name the land connection between North and South America.

    ica. _____

3. What are the islands of the Caribbean collectively

    called? _____

4. What cultural term do we use for the area made up of
    Mexico, Central America, the West Indies, and South
    America combined?

    _____

5. Name the six major bodies of water that border North
    America.

    a. _____   c. _____   e. _____

    b. _____   d. _____   f. _____

6. Name and locate Canada's major physical feature.

    _____

7. Place these major Canadian cities in order from west to
    east: Calgary, Halifax, Montreal, Ottawa, Quebec, To-
    ronto, Vancouver, Victoria, Winnipeg.

    a. _____   d. _____   g. _____

    b. _____   e. _____   h. _____

    c. _____   f. _____   i. _____

8. Name the one major Canadian city which is more than two

    hundred miles from the United States border. _____

9. Name the six regions of Canada and list the names of the provinces or territories in each.

   a. _____   c. _____   e. _____

   b. _____   d. _____   f. _____

10. Name the great river of eastern Canada. _____

11. Name the capital of Canada. _____

12. Which city is the center of the Atlantic Provinces?

_____

13. Name the second largest French–speaking city in the

   world. _____

14. Which city is the major center of English–speaking Canada?

_____

15. Name Canada's west coast gateway city. _____

16. In which city is Canada's French heritage most evi–

   dent? _____

17. Which city holds a large rodeo each year? _____

18. Which province is the heartland of Canada? _____

19. Which province is Canada's wild west? _____

20. Name Canada's two well–known Rocky Mountain Parks.

_____

21. Name the largest island off the west coast of North America.

_____

22. Name the world–famous gardens near Victoria. _____

23. Which railroad built castle–like hotels across Canada?

_____

24. Name the cities where the Frontenac and Empress Hotels are located. _____

25. Name the highest mountain in North America. _____

26. What town was once the capital of Russian America?

_____

27. Name Alaska's largest city. _____

28. What is the highlight of a cruise to Alaska? _____

29. Name the waterway taken by cruise ships along the coast of British Columbia and Alaska. _____

NOTES

CHAPTER
2

NORTH AMERICA

THE UNITED STATES—
THE EAST
AND
THE SOUTH

It is difficult to imagine a country with greater diversity in climate and landscape than the *United States of America.* Stretching from ocean to ocean across the midsection of North America with Hawaii and Alaska added for good measure, the United States' climate ranges from the Arctic to the tropics, from rain forests to deserts, from mountains to seas of grassland.

Our study of the United States will focus on the forty-eight contiguous, or connecting, mainland states, commonly, but incorrectly, referred to as the "Continental United States." After all, isn't Alaska on the continent? (We've discussed Alaska and will cover Hawaii when we reach the Pacific.)

The contiguous United States has easily definable borders: the Pacific Ocean borders the west; the Atlantic Ocean borders the east; Mexico and the Gulf of Mexico border the south; and Canada borders the north. This boundary shared by Canada and the United States extends clear across the North American continent for 4,000 miles and forms one of the world's longest borders.

The most important physical feature of the United States is its river system composed of the *Mississippi, Missouri,* and *Ohio Rivers,* flowing through and dominating the very center of the country. This massive river system runs through one of the world's richest and largest valleys (actually more an immense bowl than a valley and therefore known as the *Mississippi Bowl*) formed by the *Appalachian Mountains* to the east and the *Rocky Mountains* to the west.

East of the Appalachians is the *Atlantic Coastal Plain,* which is very narrow in the north and broadens to the south until it becomes the *Florida Peninsula.* The coastal plain is very irregular. Its many bays, peninsulas, and islands characterize the eastern coast and have created many good, natural harbors. Among its best-known features are *Long Island,* the largest island off the east coast, *Chesapeake Bay,* a long arm of the Atlantic, and the adjacent *Delmarva Peninsula* occupied by Delaware and parts of Maryland and Virginia.

West of the Rockies is the *Great Basin,* a high desert plateau. Actually, the Great Basin is another immense bowl with the Rockies forming its eastern rim and the mountains of the *Sierra Nevadas* and the *Cascades* its western rim. West of the Sierra Nevadas and Cascades are several valleys, including the great *Central Valley* of California and the *Willamette Valley* of Oregon. Along the Pacific coast is a range of mountains broken most significantly by *San Francisco Bay* and the mouth of the *Columbia River.* Lastly, *Puget Sound,* a small body of water in the northwestern corner of the country, extends around the coastal mountains and provides an outlet to the sea from deep within the state of Washington.

As with other large countries, the United States is composed of regions determined partially by its physical features, as well as such factors as history and economics. The United States is usually divided into four main regions: the *East,* sometimes called the Northeast; the *South;* the *Middle* or *Midwest;* and the *West.* However, the borders of these regions are not rigid because in this country one person's East may surely be another person's West.

THE EAST

The East generally refers to that northeastern section of the country north of the Potomac River and east of the Ohio-Pennsylvania border. This area shares a rich sense of history dating back to colonial times and has long been dominated by the chain of interlinking cities from Boston to Washington known as the *Northeastern Corridor.* Within this section of the country are six small states east of New York State called *New England.* The other eastern states are called the *Mid* or *Middle Atlantic States.*

Tucked away in the extreme northeastern corner of the country, New England is a world unto itself. Long bypassed by the mainstream of development in the United States, New England has maintained the charm and quaintness of its historic past.

Boston, at the northern end of the Northeastern Corridor, is the major center or regional hub of New England. Rich in colonial heritage, Boston offers its visitors a wonderful walking tour, known as the *Freedom Trail,* of its historic sites. Visitors also enjoy Boston's quaint neighborhoods, such as *Beacon Hill;* its cultural attractions; the campuses of many colleges and universities, including nearby *Harvard;* and especially its downtown park, the *Boston Commons.*

New England, of course, has much to offer in addition to Boston: the most beautiful autumn foliage in the United States; picturesque villages with small, cozy inns; the rugged coastline of Maine best preserved at *Acadia National Park* near Bar Harbor; skiing at Stowe, Vermont and other mountain resorts; historic sites such as *Plymouth* and *Salem* in Massa-

Portland Head, Maine

chusetts, the mansions of the turn-of-the-century-rich at *Newport,* Rhode Island; summer concerts, antique shops, festivals, and, of course, seaside holidays on *Cape Cod* and the islands of *Nantucket* and *Martha's Vineyard.*

Although Boston is the hub of New England, *New York City* is the East's, and the country's preeminent city. Affectionately known as the "Big Apple," New York is a city of constant energy and activity. Famous for its entertainment and night life, restaurants, galleries and museums, shops, sense of style, and indomitable spirit, New York is one of the world's most exciting cities.

The city has long served as the gateway to America, symbolized by the *Statue of Liberty.* Its importance, though, extends far beyond America: it is the home of the *United Nations* and the financial capital of the world, symbolized by *Wall Street.* In fact, New York City is known as a world leader in most industries, including art, publishing, fashion, entertainment, and commerce. Its many buildings, streets, and neighborhoods are famous for the trend-setting image they project: *Broadway* for theater, *Lincoln Center* for the performing arts, *Fifth Avenue* for shopping, *Madison Square Garden* for sporting events, *Metropolitan Museum of Art* for fine arts, *Greenwich Village* for bohemian life. Other city institutions are famous the world over in their own right: *Central Park,* the *Empire State Building, Times Square,* the *World Trade Center,* and *Rockefeller Center.* All of these buildings, as well as New York City's other towering skyscrapers, form Manhattan's skyline, surely the most dramatic of any city's in the world.

New York City is not only a very important destination in its own right, but it is also a major air travel center with three busy airports serving its metropolitan area: *John F. Kennedy International, LaGuardia,* and *Newark.* The city is North America's gateway to and from Europe, Africa, and parts of Asia. Moreover, it is the departure port for many cruises and most transatlantic steamship voyages.

Philadelphia, southwest of New York in the Northeastern Corridor, is famous as the *Birthplace of America.* This historic city has beautifully preserved and restored *Independence Hall,* the *Liberty Bell, Betsy Ross's home,* and many other sites associated with the founding of America.

Beyond the urbanized Northeastern Corridor, the Middle Atlantic States offer many vacation attractions in their rolling farmlands, low mountains, smaller cities and towns, and long stretches of beaches. Some of these are the *Catskill Mountain* resorts of New York State, the *Pennsylvania Dutch* country around *Lancaster,* Pennsylvania, *Atlantic City* on the beaches of New Jersey, and the *Battlefield at Gettysburg,* Pennsylvania, to name a few.

Washington, D.C. is at the southern end of the Northeastern Corridor. As the capital of the nation it has much to offer: the *Capitol* itself, the *White House,* the *Smithsonian Institution,* the *Lincoln Memorial,* the *Jefferson Memorial* to name but a few of the major public buildings and sites. Washington is especially popular with tourists in the spring when its cherry trees are in blossom. The Washington, D.C. metropolitan area is generally regarded as the border between the East and the South.

Statue of Liberty, New York City. *Photo: Courtesy New York State Department of Commerce*

THE SOUTH

The South begins somewhere in Virginia beyond the suburbs of Washington and extends all the way to the heart of Texas where the West begins. It borders the Midwest on most of its northern frontier, formed by the Ohio River and the Missouri-Arkansas and Kansas-Oklahoma borders.

More important than location, however, the southern states share a unique history and have a greater sense of unity than any of the other four regions. For many years in the past, the South truly was a nation within a nation. During the past twenty years, it has undergone remarkable growth and change and is now one of the most vital parts of the country.

Atlanta is the *hub of the South.* Having risen from the great fire of the Civil War, made so familiar to Americans by the classic novel and film, *Gone With The Wind,* Atlanta well-represents the "New South." This bustling, commercial city, termed the "New York City of the South," is the transportation center of the area and has the second busiest airport in the world.

In former days, *New Orleans* was the Queen City of the South. Today its romantic history, its French heritage, its association with jazz, the facilities offered by the *Superdome,* and the country's best-known celebration, *Mardi Gras,* make New Orleans one of America's prime tourist cities. The *French Quarter,* in the heart of the city, is the very center of the tourist's interest. New Orleans is also the major departure port for cruises up the Mississippi on Mark Twain-type steamboats, an especially popular vacation for families.

The Old South is best preserved in *Virginia.* The state's leading attraction, *Williamsburg,* is a superb restoration of the colonial capital and operates as a living museum of the seventeenth century. *Jamestown,* the first permanent English settlement in the New World, and *Yorktown,* where American independence was won, are both nearby. Washington's *Mount Vernon* and Jefferson's *Monticello* are the two most famous historic homes preserved in the state.

In the western part of Virginia, the *Blue Ridge Parkway* winds its way through the mountains and provides beautiful views of the *Shenandoah Valley.* The southern mountains, the Appalachians, are best seen, however, in *Great Smoky Mountains National Park* in North Carolina and Tennessee.

While Virginia basks in the sun of the Old South, *Florida* sparkles in the sun of the New South and has the greatest tourism industry in the region. The beaches at *Miami Beach, Fort Lauderdale,* and other communities in the Miami area, world famous as winter resorts, attract many Europeans and South Americans as well as North Americans. Miami's airport serves as a gateway to and from Latin America, and its seaport, the world's busiest for year-round Caribbean cruises, serves thousands of passengers departing its docks each week.

Other attractions in Florida include the beach resorts on the west coast centered around the *Tampa/St. Petersburg* area, *Disney World at Orlando* in central Florida, *Everglades National Park* in the far south of the state, *St. Augustine,* America's oldest city, quaint *Key West,* the southernmost city in the contiguous United States, located at the tip of the Florida Keys, and *Cape Canaveral* where the *Kennedy Space Center* is located.

Mount Vernon, Virginia. *Photo: Courtesy Virginia Division of Tourism*

Our list of attractions in the southern states seems endless. There are the restored colonial sections of *Charleston, South Carolina,* and *Savannah, Georgia,* the restored mansions of *Natchez, Mississippi,* the *Grand Ole Opry* of country and western music fame in *Nashville, Tennessee,* the beautiful bluegrass horse-raising country of Kentucky and America's favorite race, the *Kentucky Derby,* in the *Louisville, Kentucky* area, many battlefields, seaside resorts, spas, gardens, and festivals. But, keeping with our national tradition, it is time to move west, and in the South, that means Texas.

Texas is a hybrid state where the South ends and the West begins. Moreover, it has its own special history as an independent nation before joining the United States. *Houston* is its largest city and draws many conventions and sports events to its *Astrodome;* but *Dallas* is the air transportation hub of Texas and is one of the major air centers in the United States. The best-known tourist attraction in Texas, though, is in *San Antonio*—the *Alamo.* San Antonio is also known for its *Paseo Del Rio* or River Walk in its downtown area. Here a pedestrian walk winds around both sides of a narrow river through restaurants and shops, complete with music, flowers, and boats. San Antonio is truly a colorful bit of Spanish Mexico in the United States.

Two more regions of the United States beckon us; so we'll continue through the Midwest and the West in our next chapter.

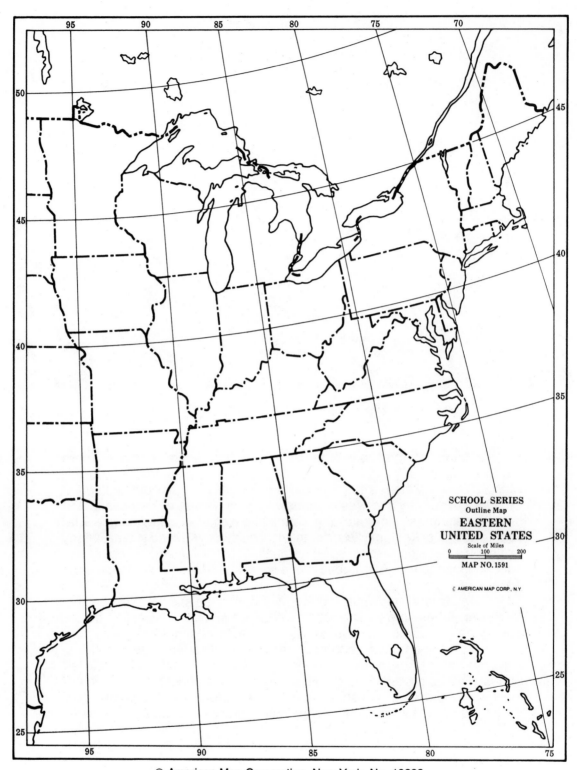

SCHOOL SERIES
Outline Map
EASTERN
UNITED STATES
Scale of Miles
0   100   200
MAP NO. 1591

© AMERICAN MAP CORP., N.Y.

NORTH AMERICA

The United States—the East and the South

1. Name the two subdivisions of the East. _____

2. What is the chain of cities in the East called?

   _____

3. Name the preeminent city of North America. _____

4. Name the hub of New England. _____

5. Which city is famous as the Birthplace of America?

   _____

6. Name the ''New York City of the South.'' _____

7. Which city is a bit of Spanish Mexico in the United States?

   _____

8. Name the air transportation hub of Texas. _____

9. Which southern state has the greatest tourism indus-

   try? _____

10. In which city would you find each of the following?

    a. Paseo Del Rio _____

    b. The French Quarter _____

    c. Freedom Trail _____

    d. Lincoln Memorial _____

    e. Lincoln Center _____

    f. Greenwich Village _____

    g. Independence Hall _____

    h. Disney World _____

    i. The Alamo _____

    j. Astrodome _____

    k. Superdome _____

    l. Grand Ole Opry _____

11. Name the three airports in the New York City area.

_____

_____

_____

12. Washington, D.C. is at the border of which two regions?

_____

_____

13. Name the United States air gateway to Europe and Africa. _____

14. Name the United States air gateway to Latin America.

_____

15. Which city is the world's busiest cruise departure port?

_____

16. Name North America's oldest city. _____

17. In which state or states would you find the following?

    a. Cape Canaveral _____

    b. Blue Ridge Parkway _____

    c. Cape Cod _____

    d. Nantucket _____

    e. Catskill Mountains _____

    f. Mount Vernon and Monticello _____

    g. Williamsburg, Jamestown and Yorktown _____

18. What is the best-known celebration in the United States?

_____

19. Name the national park at the southern tip of Florida.

_____

20. Which national park protects a portion of the coast of

Maine? _____

21. Which national park protects the beauty of the Appalachians?

_____

22. Name the southernmost city in the contiguous United

States. _____

23. In which state does the South meet the West?

_____

24. In which state is the Old South best preserved?

_____

25. Going from northeast to southwest, place the following
cities in order: Baltimore, Boston, Hartford, Newark,
New York, Philadelphia, Portland, Providence, Trenton,
Wilmington, Washington.

a. _____   g. _____

b. _____   h. _____

c. _____   i. _____

d. _____   j. _____

e. _____   k. _____

f. _____

NOTES

CHAPTER
3
NORTH AMERICA

THE UNITED STATES—
THE MIDWEST
AND
THE WEST

THE MIDWEST

The Midwest occupies the central part of the country, above the South, from the Pennsylvania-Ohio border to the Plains States, the tier of states above Texas and Oklahoma, including Kansas, Nebraska, and the Dakotas. The Great Lakes are the very core of the region and the Midwest states which border directly on the lakes are sometimes subdivided as the *Great Lakes States.* The region is the workhorse of the country both in its industrial centers like Detroit and its famed agricultural areas such as the corn belt. The size of its lakes and its industrial and agricultural strength have helped to set the tone for this region as a land of giants.

To start with, its hub, *Chicago,* is the most important air transportation center in the country because its airport, *O'Hare Field,* is the world's busiest. Moreover, as the center of this productive area, Chicago is a giant among cities and it has the buildings to prove it: the Standard Oil Company has eighty stories, the John Hancock Center has ninety-five stories, and the *Sears Tower* has one hundred and ten stories making it the world's tallest building. Chicago's well-known "Loop," in its downtown area, includes restaurants, theaters, and many stores. Best known among the stores is the huge *Marshall Field's* department store, which is so large it offers guided tours.

Resorts abound on the Great Lakes and on many of the smaller lakes as well. *Mackinac Island* in Lake Huron is the most famous of these resorts. Access to this island is by ferryboat. It allows no cars; visitors and residents travel by bicycle and horse and carriage. The island is especially famous for the *Grand Hotel,* appropriately, the largest summer hotel in the world.

Even the major tourist sites of the area are the stuff of giants. Best known among them is a symbol of the United States herself—*Mt. Rushmore, South Dakota.* There, giant-sized heads of four presidents—Washington, Jefferson, Lincoln, and Theodore Roosevelt—have been carved out of granite.

Another man-made attraction on the giant scale is the 634-foot-high *Gateway Arch* on the banks of the Mississippi at *St. Louis, Missouri.* The arch, which has small trains to carry visitors to an observation tower, was built to symbolize the role St. Louis played as the gateway to the west.

THE WEST

Today the *West* as a region of the country is some distance west of St. Louis. Indeed, it would be easy to say that the west begins at Denver or, in any case, west of the tier of Plains States that mark the westernmost states of the Midwest (Kansas to North Dakota). From the western borders of those states, the West sweeps across the rest of the United States. Although the West is the largest in size of the four regions, it is, with the great exception of California, the most sparsely populated.

Obviously, such a large region would have several subdivisions based on geography, history, and culture. The West is commonly divided into these four areas: the *Mountain West,* the *Pacific Northwest,* the *Southwest,* and the *West Coast.*

Gateway Arch, St. Louis, Missouri. *Photo: Courtesy Missouri Tourism Commission*

THE MOUNTAIN WEST

The *Mountain West* is "Marlboro Country" or what some people would call the "real west," encompassing the Rocky Mountain States and the Great Basin States: Montana, Wyoming, Colorado, Utah, Idaho, and Nevada. Although Nevada is clearly a Great Basin State, its tourist attractions are so closely tied with those of California that we will discuss them within the context of the West Coast. The Mountain West is still cowboy and ranch country; however, it is undergoing rapid change and is becoming known as a source of energy with mines and rigs strewn across the range. *Denver* is the hub of the Mountain West. A major air terminal, Denver is also the gateway to many of the mountain ski resorts such as *Vail* and *Aspen, Colorado,* and *Jackson Hole, Wyoming. Salt Lake City, Utah,* also provides easy access to many of the ski resorts in the mountains. As the center of the Mormon Church, Salt Lake City offers much of historic and religious interest around its *Tabernacle Square.*

The Mountain West is an area of spectacular scenery which is protected in a number of national parks including *Glacier* in Montana, *Yellowstone* with its many natural wonders in Wyoming, *Rocky Mountain* in Colorado, and *Zion* and *Bryce Canyon* in Utah.

Swiftcurrent Lake, Glacier National Park, Montana. *Photo: Courtesy Travel Promotion Bureau, Montana Dept. of Highways*

While these great national parks draw many vacationers, others head to the Mountain States for a taste of the West—a ranch vacation, visits to restored mining towns, trips on narrow gauge railroads and, of course, attending rodeo. The Silverton-Durango train which carries passengers on a scenic route through southwestern Colorado is the best known of the narrow gauge trains, and the rodeo at Cheyenne's annual *Frontier Days* has a strong claim as the country's best rodeo.

THE PACIFIC NORTHWEST

The *Pacific Northwest* is also known for its spectacular scenery. This region occupies the northwestern section of the country, and includes Washington, Oregon and the most northern sections of California. The Pacific Northwest evokes soft, misty, green images of rain forests, redwoods, lumber towns, rugged seacoasts, and fishing villages. Its hub is *Seattle,* located on *Puget Sound,* which gives the city a maritime air.

Seattle seems to nestle at the foot of *Mt. Rainier,* one of the country's most beautiful peaks. This area is noted for beautiful mountains including *Mt. Hood* in Oregon which is especially popular with skiers. Best known of all these mountains though is *Mt. St. Helens.*

Seattle also provides easy access to the *Olympic Peninsula,* located at the extreme northwest of the contiguous United States. There *Olympic National Park* protects the rain forest and some of our most unique mountain and seacoast scenery. The Pacific Northwest is actually as well known for its coast as for its mountains; especially noteworthy is the magnificent, rugged coast of Oregon.

THE SOUTHWEST

In contrast to the soft, misty green images of the Pacific Northwest, the *Southwest,* centered in Arizona and New Mexico, explodes with the vivid colors of the sun and the cultures of the Indians, Spaniards, and Mexicans. With its open spaces, wonderful climate, and easy access to the cities of Texas and Southern California, the Southwest is the fastest growing region of the country.

Its hub is *Phoenix, Arizona,* which enjoys the resort city of Scottsdale right next door. Two smaller cities in neighboring New Mexico also draw many tourists. The architecture and streets of *Santa Fe* and nearby *Taos* reflect the Indian, Spanish, and Mexican influences of the region. Moreover, Santa Fe is an important summer opera center and Taos is a winter ski resort. Both cities are also well-known artists' centers.

The highlight of the Southwest, though, is *Grand Canyon National Park, Arizona.* Quite simply, it is one of the greatest natural spectacles in the world and can be enjoyed in many ways—from mule-train trips to scenic flights.

THE WEST COAST

Many visitors to the Southwest also travel to the *West Coast* which simply means California from the San Francisco Bay area south. This area is magic for most North Americans and for much of the rest of the world as well.

The West Coast revolves around two major centers—Los Angeles and San Francisco. *Los Angeles* is the larger and more important of the two cities and is one of the major air transportation centers of the United States. It is an important departure point for cruises to Mexico and flights to much of Latin America. It is also the gateway to and from Asia and the Pacific. Los Angeles is also a major destination in its own right for business travelers and for tourists drawn by *Hollywood, Beverly Hills,* nearby *Disneyland* in Anaheim, and the beaches and other attractions of Southern California.

Among these is the city of *San Diego,* which has developed into a beautiful beach and boating resort and has one of the world's finest zoos. In contrast to San Diego's oceanfront setting but sharing its sunshine is *Palm Springs,* a popular desert resort renowned for its tennis courts, golf courses, and spas. Visitors to Southern California usually include a trip to *Las Vegas, Nevada,* too. This desert city is one of the entertainment and gambling capitals of the world, and, truly, one of the wonders of the modern age. Its main street, called "the Strip," lined with hotels and casinos, lures visitors from around the world.

San Francisco, 400 miles to the north of Los Angeles, is the center of the northern section of the West Coast. Commonly called "America's

Sleeping Beauty Castle, Disneyland, Anaheim, California. *Photo: Courtesy The Walt Disney Company,* © *1986*

favorite city," San Francisco combines its all-American history as the California-gold-rush-city and onetime Queen of the West with strong European and Oriental influences, particularly Italian and Chinese, which makes it one of the most cosmopolitan cities in the world.

San Francisco offers visitors spectacular views of its natural setting, the finest collection of Victorian homes in the country, its colorful *Chinatown,* the *Alcatraz Penitentiary, Fisherman's Wharf,* and, the two universal symbols of the city, the *Golden Gate Bridge* and the cable cars.

Ranging out from San Francisco, tourists can journey to the nearby *Wine Country* reminiscent of France and Italy, and to the year-round resorts of *Lake Tahoe,* on the border with Nevada. *Yosemite,* the most beautiful of the national parks for many people, located in the Sierras, is also included on trips to Northern California.

Between Los Angeles and San Francisco, visitors can enjoy a drive up California Highway #1 which winds its way along the edge of mountains high above the Pacific in an area known as *Big Sur.* On this route, they can visit *San Simeon,* the palatial home of the Hearst family, the charming town of *Carmel,* and the nearby *Pebble Beach Golf Course.*

Having toured Anglo-America, it is time to turn south of the border to enjoy the bright colors and sunshine of Mexico, the West Indies, and the other areas of Latin America.

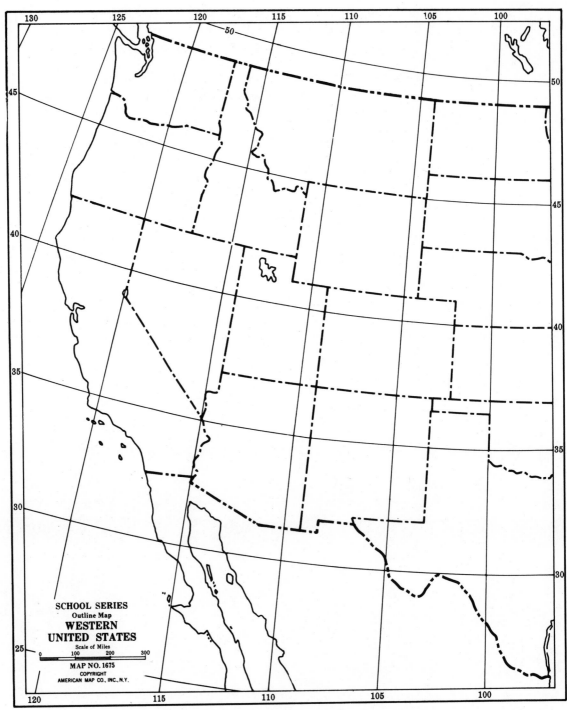

SCHOOL SERIES
Outline Map
**WESTERN
UNITED STATES**
Scale of Miles
0   100   200   300
MAP NO. 1675
COPYRIGHT
AMERICAN MAP CO., INC., N.Y.

© American Map Corporation, New York, No. 19222

North America

The United States—the Midwest and the West

1. Name and locate the world's busiest airport.

   _____

2. Name and locate the world's largest summer hotel.

   _____

3. How many Great Lakes are there? Which, if any, of the Great Lakes is entirely within the United States?

   _____

   _____

   _____

4. At what city can we say the West begins today?

   _____

5. Name the four subdivisions of the West.

   _____

   _____

6. Name the most important city in each subdivision.

   _____

   _____

7. In which subdivision is each of these states?

   a. Wyoming _____   c. Oregon _____

   b. Colorado _____   d. Arizona _____

8. What is the only state in the West Coast subdivision?

   _____

9. Name the peninsula in the far northwestern corner of

   the United States. _____

10. Name the mountain overlooking Seattle.

    _____

11. Name the Southwestern city with a summer opera pro-

    gram. _____

12. What is the highlight of the Southwest?

    _____

13. Name the air transportation gateway to Asia and the Pa-

    cific. _____

14. Which West Coast city has one of the world's finest

    zoos? _____

15. Which city is commonly called ''America's favorite

    city''? _____

16. Name the Hearst palatial home. _____

17. Name the lake at the center of the California-Nevada

    resort area. _____

18. What and/or where are the following?

    a. The Strip _____

    b. Golden Gate Bridge _____

    c. Gateway Arch _____

    d. Marshall Field's _____

    e. The Loop _____

    f. Alcatraz _____

    g. Tabernacle Square _____

    h. Sears Tower _____

19. In which state or states would you find the following National Parks?

a. Yosemite _____   d. Grand Canyon _____

b. Glacier _____   e. Rocky Mountain _____

c. Bryce Canyon _____   f. Olympic _____

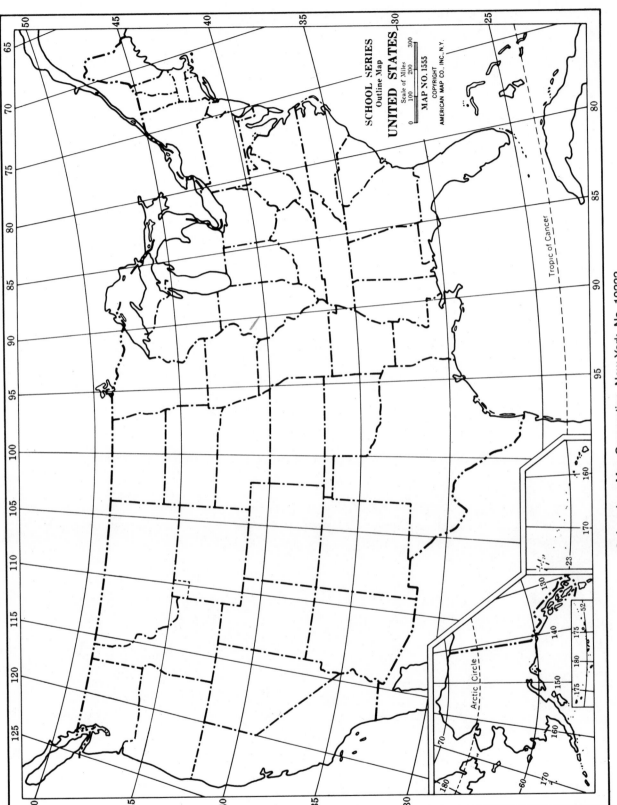

SCHOOL SERIES
Outline Map
UNITED STATES
Scale of Miles
0   100   200   300
MAP NO. 1555
COPYRIGHT
AMERICAN MAP CO. INC., N.Y.

Tropic of Cancer

Arctic Circle

North America
The United States—a Review

1. What is meant by the contiguous United States?

   _____

2. Give the borders of the contiguous United States.

   _____

   _____

3. Name the most important physical feature in the United
   States. _____

   _____

4. Name the mountains which form the western and eastern
   rim of the great Mississippi Bowl. _____

5. Name the largest island off the east coast of the
   United States. _____

6. Name the Great Lakes. _____

   _____

7. Name the desert area between the Rocky Mountains and
   the Sierra Nevada Mountains. _____

8. Name the body of sea water which cuts deep into Washington State. _____

9. Name the bay where the coastal range of mountains on
   the west coast is broken. _____

10. Name the state that borders both the country of Mexico
    and the Gulf of Mexico. _____

11. How many states border the Gulf of Mexico? _____

12. Name the state which borders both the Gulf of Mexico and the Atlantic Ocean. _____

13. What is the term given the six small states located in the extreme northeastern corner of the United States?

_____

14. How many states border the Great Lakes? _____

15. How many contiguous states border the Pacific Ocean?

_____

16. Name the state which borders Mexico and the Pacific Ocean. _____

17. How many states border Mexico? _____

18. Which river is the border between the United States and Mexico? _____

19. Name the two states which border more states than any others. How many states does each of them border?

_____

_____

20. Name the one state which borders only one other state.

_____

Which state does it border? _____

21. Name four states which each border only two other states. _____

_____

22. Name the four regions of the United States.

_____

_____

23. Which river is the border between the South and the

    Midwest? _____

24. Name two states which border no other states.

    _____

25. What is the least number of states you could pass
    through traveling from the Pacific Ocean to the Atlan-
    tic Ocean across the contiguous United States?

    _____

NOTES

CHAPTER
4
NORTH AMERICA

MEXICO,
CENTRAL AMERICA,
AND
THE WEST INDIES

MEXICO

A cosmopolitan, international capital, beautiful beaches, mysterious archaeological sites, quaint colonial towns, sunshine, bright colors, and traditional handicrafts—Mexico has them all. Its proximity as the closest non-English speaking culture to the United States, its relatively reasonable cost, good air service to its key tourist destinations, and aggressive advertising campaigns by government and private industry explain Mexico's immense drawing power as a tourist destination.

Mexico arches southeast from the American Southwestern States to Guatemala and Belize in Central America. Its western shore is bordered by the *Pacific Ocean,* its eastern by the *Gulf of Mexico* and the *Caribbean Sea.* The country is, for the most part, a high, rolling plateau between two ranges of mountains, the *Sierra Madre Occidental* on the west and the *Sierra Madre Oriental* on the east. There is a thin coastal strip on each side of the country.

Mexico also has two peninsulas. *Baja California* extends from its northwest corner and is separated from the mainland by the *Gulf of California* (sometimes called the Sea of Cortez). The *Yucatán* extends from the southeastern corner of the country and separates the Gulf of Mexico from the Caribbean Sea.

For tourist purposes, the country is usually divided into five zones: *Mexico City,* the *Mexican Riviera, Baja California,* the *Yucatán,* and the *Colonial Route.*

MEXICO CITY

Mexico City is the very heart of Mexico. It is the largest city in the country as well as one of the largest in the world. Built on the same site as the cities of earlier Indian civilizations, Mexico City stands at 7,000 feet—one of the highest cities in the world. The altitude, combined with the city's southern latitude, at least by North American standards, gives Mexico City a year-round mild climate.

This busy, fast-growing city has many attractions for tourists. The *Zona Rosa* is the primary tourist area with airline offices, many fine hotels and restaurants, smart shops and boutiques, discos, and night clubs. The *Zocalo* is the main plaza of the city. It is surrounded by many important public buildings including the cathedral, town hall, National Palace, and Supreme Court as well as its famous Aztec ruins. The university area with famous murals painted on building walls also draws many visitors.

Sundays are an especially popular and pleasant day to be in Mexico City, for visitors have a wide choice of activities. They can enjoy a performance of the *Ballet Folklorico,* Mexico's internationally famed dance troupe; go to a bullfight (from November to April); drift through the floating *Gardens of Xochimilco* on a flower-adorned boat; or wander through *Chapultepec Park* sharing the day with Mexican families, entertainers, and colorful vendors, who bring a sense of vitality to the lawns, groves, gardens, paths, and boating lake of this vast park.

For those interested in Mexico's pre-colonial past, the city offers the *National Museum of Anthropology,* where the displays rival the artifacts, and the nearby *Pyramids of Teotihuacan,* a reminder of the country's ancient Indian civilization.

Teotihuacan pyramids and ruins. *Photo: Courtesy FONATUR*

Two nearby towns are often included in a trip to Mexico City: *Cuernavaca,* a fashionable resort especially famous for its gardens and villas, and *Taxco,* one of the jewels of Mexico. This old silver mining town, built on a beautiful hillside, is just what most people expect of "Old Mexico." As a national landmark, the hilly cobblestone streets, whitewashed houses, red tile roofs, garden walls, bright bougainvillea, cathedral, and central plaza, which all together give the town its charm, will be protected and preserved.

MEXICAN RIVIERA

Many tourists are more interested in the new Mexico of sparkling beaches and dazzling resorts, and, in all but the hottest summer months, they flock to the sunny seaside resorts. If Mexico City is the heart of Mexico, these beach resorts are the outermost fringe. The favorite resort area is along the west coast of the mainland of Mexico, called the *Mexican Riviera.*

Acapulco is the most sophisticated of the resort cities along the Riviera. It not only offers fine beaches and luxury hotels in a beautiful setting, but also excellent shopping, restaurants, discos, night clubs, and the famous cliff divers. Beginning each evening at 9:30 and continuing hourly until after midnight, divers, carrying torches, leap 136 feet from the cliffs into the sea below.

Puerto Vallarta, also beautifully located, has maintained more of its distinctive Mexican flavor than has Acapulco. Many prefer this city for its greater informality and more relaxed ambience. Visitors to Puerto Vallarta can enjoy a two-hour cruise along cliffs and hidden sandy coves to the island of *Yelapa.* The island recalls the South Seas with its palm-lined beach and several open-air seafood restaurants.

Zihuatanejo/Ixtapa is the newest resort area on the Riviera. "Z," as its devotees call it, is a wonderfully untouched fishing village. Nearby *Ixtapa* is a resort area of large modern hotels. Both have excellent beaches. Together, they are an ideal combination. Other resort cities on the Riviera include *Mazatlán,* famous for its sports fishing; *Manzanillo,* known for its two self-contained resorts; *Las Hadas,* patterned after a Moorish fantasy; and *Club Maeva.*

BAJA CALIFORNIA

Baja California, until recently one of the most remote areas of North America, is being developed as a tourist destination. It is still, however, a nature lover's paradise. Most of the development is taking place at *La Paz* and *Cabo San Lucas.* While the amenities and ambience here are not the same as on the Riviera, the towns do provide excellent vacation facilities.

YUCATÁN

In contrast, Mexico's other peninsula, the *Yucatán,* has highly developed tourist facilities especially in *Cancún* and *Cozumel.* The village of *Cancún* was selected, by computer, for development as a major resort. Today it is one of the most fashionable resorts in the world. *Cozumel* is an island off the coast of the Yucatán. In addition to fine hotels and beaches, both resorts offer especially good scuba diving and snorkeling in the clear waters which surround them. Indeed, *Palancar,* the coral reef near Cozumel, is ranked as one of the world's top diving spots.

Yucatán, however, offers much more than beach resorts to tourists. As the home of the ancient Mayas, the Yucatán holds some of the world's richest archaeological treasures. Best known among the Mayan ruins are those of *Chichén Itzá,* one of the world's most extensive archaeological sites, and nearby *Uxmal.*

COLONIAL ROUTE

The *Colonial Route* of Mexico, north of Mexico City, is rich in Spanish colonial history and could well be recommended to clients who are interested in art, architecture, history, and culture. *Guadalajara* is the best known of the cities on this route. It has some striking examples of colonial architecture, but is noted for its outdoor market, said to be the most extensive in Mexico. *Guanajuato* is a city rich in history, silver, and charm, and is today one of the most interesting and colorful cities in Mexico.

Morelia exemplifies Spanish colonial architecture and town planning so well that it looks like a European city. It is said to have the most beautiful cathedral in Mexico. *San Miguel de Allende* is another jewel of Mexico. A writers' and artists' colony, it is well known for its handicrafts and its language institute, attended by many Americans and Canadians.

CENTRAL AMERICA

Seven small countries extend from Mexico to South America. They are collectively called Central America. Although some of these countries have tried to promote tourism, the political instability of the region has kept many tourists away from the area. The popular tourist destination in the area is the *Panama Canal,* which is usually visited by cruise passengers traveling from the Caribbean to the Pacific or vice versa.

WEST INDIES

In contrast to Central America, the West Indies are among the most attractive destinations for North American tourists. The West Indies extend southeast from Florida like giant stepping-stones to South America. They are almost synonymous with the Caribbean Sea which they actually form by separating its waters from the Atlantic Ocean. Despite the difference in their colonial backgrounds, these islands share a natural heritage of sun, sand, surf, and agriculture—particularly sugar—and a mixture of African and European cultures.

Tourism is vital to the economies of the West Indies and most of the island-nations actively promote it. While cruise ship stopovers are the most common way for Americans and Canadians to visit the islands, many North Americans also enjoy resort vacations there. Then, too, there is a growing interest in private yacht rentals especially in the more remote islands.

Most of the cruises to the Caribbean depart from the Miami area, including Fort Lauderdale and Port Everglades. Since most are one-week

Shaw Park Gardens, Ocho Rios, Jamaica. *Photo: Courtesy Jamaica Tourist Board*

cruises, their itinerary is limited by the constraints of time. Among the most popular of the ports of call are two United States territories in the area, *Puerto Rico* and *St. Thomas.*

San Juan, Puerto Rico, is known for its charming *Old Town* and historic fortress *El Morro,* its Las Vegas-style nightlife, its gambling, and racetrack. It also is an important cruise departure port and provides a good alternative base for one-week cruises—and a subsequent itinerary—for those cruise enthusiasts who have already taken a one-week cruise from Miami and want to sail to the more distant islands.

The islands of *St. Thomas, St. John,* and *St. Croix* make up the *U.S. Virgin Islands,* purchased by the United States from Denmark during World War I. The islands still ring with Danish names such as Christiansted. The major town on St. Thomas is *Charlotte Amalie,* a duty-free port, which is a great favorite with cruise ship passengers. Nearby St. John is well known for its Caribbean National Park.

Resort vacationers are drawn to the beautiful beaches and mountains of many of the islands. *Montego Bay* and *Ocho Rios,* both in *Jamaica,* are two of the most popular resort areas. Some visitors prefer the French ambience of *Martinique* with its sampling of French culture, cuisine, and history including the birthplace of the Empress Josephine, wife of Napoleon, and the ruins of St. Pierre, known as the "New Pompeii," where 30,000 people died in a volcanic eruption in 1902.

The Bahamas, islands off the east coast of Florida in the Atlantic, and *Bermuda,* east of the Carolinas in the Atlantic, are often grouped with the West Indies. Both have British backgrounds, share many of the features of the Caribbean islands, and enjoy a brisk tourist trade. Each serves as a destination for shorter cruises from the mainland and each draws many visitors to its resorts.

San Jose Church, Old San Juan, Puerto Rico. *Photo: Courtesy Puerto Rico Tourism News Bureau*

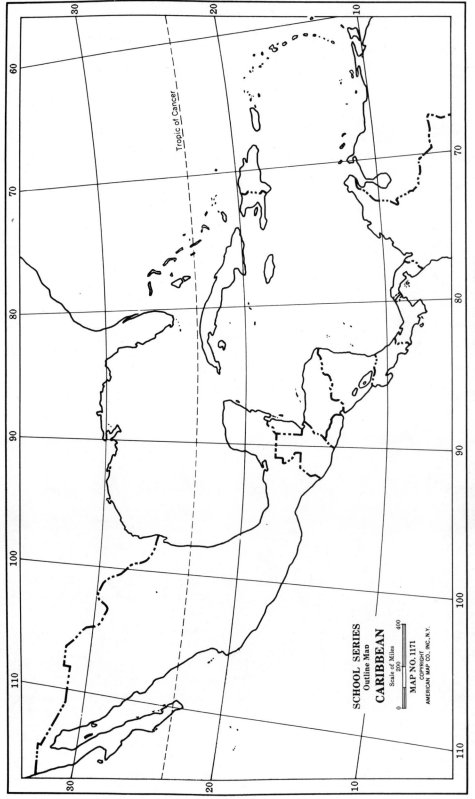

SCHOOL SERIES
Outline Map
CARIBBEAN

Scale of Miles

0   200   400

MAP NO. 1171
COPYRIGHT
AMERICAN MAP CO., INC., N.Y.

Tropic of Cancer

NORTH AMERICA

Mexico, Central America, and the West Indies

1. What river provides a border between the United States

and Mexico? _____

2. The shores of Mexico are bordered by two Gulfs, a Sea,
and an Ocean. Name them.

_____

_____

_____

_____

3. Name the two Mexican peninsulas.

_____

_____

4. Name the seven Central American nations.

_____

_____

_____

5. What are the two Atlantic Ocean island groups often

grouped with the West Indies? _____

6. Name the five tourist regions of Mexico.

_____

_____

_____

7. Name four Sunday events in Mexico City.

_____

_____

_____

_____

8. Name five resort cities on the Mexican Riviera.

_____

_____

_____

_____

_____

9. Which is the most sophisticated of the Mexican Riviera

resorts? _____

10. Which of the Mexican Riviera resorts is especially

well known for its fishing? _____

11. Name the resort island off the coast of the Yucatán.

_____

12. Name the silver mining town which is a national land-

mark in Mexico. _____

13. What and where are the following:

a. Zona Rosa _____   f. La Paz _____

b. Yelapa _____   g. Charlotte Amalie

c. Chichen Itza _____   _____

d. Zocalo _____   h. Ballet Folklorico

e. Ocho Rios _____   _____

    i. El Morro _____   k. San Miguel de Allende

    j. Morelia _____   _____

                        l. New Pompeii _____

14. What is the most important tourist feature in Central America? _____

15. Name the two United States territories in the West Indies. _____

16. Which European country is the ''mother country'' of Martinique? _____

17. Which of the West Indian Islands has a United States national park? _____

18. Name the West Indian city with Las Vegas–style gambling and entertainment. _____

19. In which Caribbean or Atlantic island–country are each of the following cities located?

    a. San Juan _____   g. Bridgetown _____

    b. Kingston _____   h. Charlotte Amalie

    c. Port-au-Prince   _____

       _____   i. Santo Domingo _____

    d. Port of Spain   j. Pointe-à-Pitre

       _____   _____

    e. Fort de France   k. Havana _____

       _____   l. Nassau _____

    f. Hamilton _____

20. Starting at the Florida end of the West Indies and moving towards the South American end, place the following countries in order: Barbados, Dominican Republic, Cayman Islands, Guadeloupe, Haiti, Jamaica, Martinique, Puerto Rico, Trinidad and Tobago, U.S. Virgin Islands.

a. _____   f. _____

b. _____   g. _____

c. _____   h. _____

d. _____   i. _____

e. _____   j. _____

CHAPTER
5

SOUTH
AMERICA

South America is one of the world's most exciting places! Imagine rhythmic samba dancers on the dazzling beaches of Rio de Janeiro, stoic Indians tending herds of llamas in the high Andes, and dugout canoes drifting down the crocodile-filled Amazon through a dense tropical jungle. Enticing images such as these are, in fact, clues to the range of the continent's landscapes and cultures—from snow-covered peaks to rain forests, from native Indians to transplanted Europeans.

South America is dominated by her two outstanding physical features: the *Andes Mountains* and the *Amazon River.* The Andes, one of the great mountain ranges of the world, form a spinal column down the continent's west coast. They reach their greatest height in the southwest on the Argentina-Chile border, where *Mt. Aconcagua* towers 22,835 feet, making it the highest point in the Western Hemisphere. The Amazon River is one of the longest river systems in the world, and its valley contains the world's largest rain forest. The rain forest and its wildlife can be enjoyed comfortably by tourists from cruise ships which ply the river.

BRAZIL

Both the river and the rain forest are almost entirely within the vast country of Brazil. Not only is Brazil the largest and most populous country in South America, it is the fifth largest and seventh most populous country in the world. It occupies half of South America and borders every other country on the continent, but two. As a Portuguese-speaking nation, it differs significantly from the rest of South America which is almost entirely Spanish-speaking. Furthermore, its culture is an interesting blend not only of its European and Indian pasts, but of its African heritage as well.

Brazil is also the most popular tourist destination in South America, primarily because *Rio de Janeiro* draws many visitors. One of the best-known and loved cities in the world, Rio is famous for its magnificent setting, its superb beaches, such as *Ipanema* and *Copacabana,* and, of course, its Carnival. Rio's wonderful setting can best be seen from *Sugar Loaf Mountain,* accessible by cable car, or *Corcovado Mountain* with its 130-foot statue, *Christ the Redeemer.* Carnival brings thousands of tourists to Brazil to join in the festivities which occur before the arrival of Lent. This feast is similar to the Mardi Gras that takes place in New Orleans, and, along with Oktoberfest in Munich, Carnival is probably one of the two most famous celebrations in the world. Another popular destination among tourists is *Iguazú Falls,* one of the world's greatest falls, located near the border of Paraguay, Argentina, and Brazil.

While vacationers flock to Rio and the Iguazú Falls, business travelers head for *São Paulo,* a commercial and industrial center which is not only Brazil's largest city but also one of the world's largest.

SOUTH AMERICAN CARIBBEAN COUNTRIES

The countries north of Brazil—*French Guiana, Suriname, Guyana, Venezuela,* and *Colombia*—are often referred to as the South American Caribbean countries. Although only the last two actually border on the Caribbean, all share a common geography and a common history associated

Christ the Redeemer, Rio de Janeiro, Brazil. *Photo: Courtesy Varig Airlines*

with the Caribbean. Oil-rich *Venezuela* and its modern capital, *Caracas,* are the best-known country and city in this region. Venezuela is further linked to the Caribbean Islands through its seaside resorts and cruise ship stops such as *Maracaibo* and *La Guaira.*

French Guiana is best known for *Devil's Island,* a penal colony just off its coast. One of the most notorious prisons in the world, Devil's Island gained considerable fame through the novel and film *Papillon.* It was closed as a prison in 1945 and is now open to tourists.

ANDEAN COUNTRIES

To the west of Brazil are the *Andean countries—Ecuador, Peru, Bolivia,* and *Paraguay.* Although Paraguay does not share the Andean landscape of the other three, it is associated with them because, like them, its people and culture are predominantly Indian.

Peru is another favorite country of tourists in South America primarily because of the remnants of ancient Inca culture there. The Spanish colonial city of *Lima* is the gateway to Peru. Although several fine museums are located in Lima, *Cuzco* and *Machu Picchu* are the major centers for visiting the world of the Incas. *Cuzco,* once the capital of the Inca Empire, was built over by the Spanish conquistadors. Both cultures are very evident in this colorful city which serves as the gateway to *Machu Picchu,* "The Lost City of the Incas." This magnificent ruin of an Incan city, never discovered by the Spanish, is considered by many to be the foremost archaeological site in South America. Machu Picchu blends its mysterious past and spectacular location high in the mountains so well that it is often the highlight of a trip to South America.

Peru is also well known for the huge and mysterious land drawings at *Nazca.* Predating the airplane by centuries, these immense pictures can only be seen from the air!

Many tourists to Peru also visit the Amazon River and rain forest at *Iquitos,* located in the Peruvian jungle near the headwaters of the river. Peru, along with Bolivia, borders *Lake Titicaca,* the highest navigable lake in the world, famous for its reed boats.

In the Andean area, *Quito,* capital of Ecuador, is another favorite tourist site. It is a beautifully preserved and charming Spanish colonial city amidst splendid mountain scenery.

SOUTHERN COUNTRIES

The countries south of Brazil—*Uruguay, Argentina,* and *Chile*—are the most European of all the South American countries. *Buenos Aires,* capital of Argentina, is the most important city in this region. One of the largest cities on the continent, it is often called the "Paris of South America" for the beauty of its wide boulevards, its shops and restaurants, the sophistication of its people, and the European flavor of its life. The *Boca,* sometimes called "Little Italy," is a particularly lively waterfront district of the city which draws many to its fine restaurants.

Argentina is especially known for its great grasslands called the *Pampas.* Cattle are raised there, herded by the Argentine cowboy, the *gaucho.* Tourists often visit an estancia, or cattle ranch, when touring this area.

Both Argentina and neighboring Uruguay are famous for their beautiful beaches. *Mar del Plata* is the leading beach resort area of Argentina and boasts the world's largest casino. *Punta del Este* is Uruguay's leading beach resort and is an international destination. Remember that we are now in the Southern Hemisphere, so the beach season is the reverse of ours.

Argentina and Chile share a magnificent alpine area with beautiful lakes, skiing facilities, and fine resorts. The best-known center for this area is *San Carlos de Bariloche,* Argentina.

South America ends at a point in the south called *Cape Horn.* The passage through this wild and remote area is called the *Straits of Magellan.* This beautiful wilderness is one of the highlights of a cruise around South America.

The Galápagos Islands, Ecuador, and Easter Island, Chile, off the west coast, are well known to world travelers. The *Galápagos Islands* with their rare plants, birds, and animals are of special interest to naturalists. *Easter Island* is famous for its gigantic stone statues of unknown origin.

South America with its Latin heritage is considered, along with Mexico, Central America, and the West Indies, a part of *Latin America.* To help

Mysterious Stone Statues of Easter Island, Chile. *Photo: Courtesy Jim Woodman*

yourself remember its regions, recall your *ABC'S,* "A" for Andean countries, "B" for Brazil, "C" for Caribbean countries, and "S" for the southern countries.

We have now completed our tour of the New World or Western Hemisphere. Next we shall turn our attention to the Old World, to Europe, to gain an appreciation of the many attractions which make it a favorite destination for peoples all over the world.

Medellin Monumento, Colombia. *Photo: Courtesy Colombian Government Tourist Office*

Equator

Tropic of Capricorn

Cleartype®

Outline Map
SOUTH AMERICA

AMC

0   200   400   600   800 Miles

0   400   800 Kilometers

S-MR-R

Map No: School Series 1531

© American Map Corporation, New York, No. 19222

SOUTH AMERICA

1. Name the largest and most populous country in South America.

   _____

2. Name the only countries which do <u>not</u> border that
   country. _____

3. Name the countries which are included in South America's Caribbean area. _____

   _____

4. Which country is oil rich? _____

5. Which country borders on both the Caribbean Sea and
   Pacific Ocean? _____

6. Name the countries which are included in the Andean area.

   _____

   _____

7. Which three countries are the most European?

   _____

8. Name the two countries which have no sea coast.

   _____

9. What is the tip of South America called? _____

10. Name the Straits which provide a passage through the tip.

   _____

11. What is the name of South America's mighty mountain range?

   _____

12. Name and locate the highest peak in the Western Hemisphere.

   _____

13. Name and locate the highest navigable lake in the world.

   _____

14. What is the major river on this continent? _____

15. Near the borders of which three countries is the

   Iguazú Falls? _____

16. Which city is the ''Paris of South America''?

   _____

17. Name the best known center of the Argentine–Chilean alpine

   area. _____

18. Name the ''Lost City of the Incas.'' _____

19. Which islands are of special interest to naturalists?

   To which country do they belong? _____

   _____

20. Which city is the most popular tourist city in South

   America? _____

21. Where is the world's largest casino? _____

22. Which island is famous for its gigantic stone statues?

   To which country does this island belong? _____

   _____

23. Name the commercial and industrial center of Brazil.

   _____

24. Name the capital of Brazil. _____

25. Name the most famous celebration in South America. In

   which country does it take place? _____

   _____

26. What and in what country can you find:

    a. The gaucho _____   e. Nazca _____

    b. Ipanema and   f. Boca _____

       Copacabana _____   g. The Pampas _____

    c. Corcovado Mountain   h. Cuzco _____

       _____   i. Punta del Este

    d. Devil's Island   _____

       _____

27. Starting at Caracas, place the following cities in
    clockwise order: Buenos Aires, Bogota, Caracas, La Paz,
    Lima, Quito, Rio de Janeiro, San Carlos de Bariloche,
    Santiago, São Paulo.

    a. _____   f. _____

    b. _____   g. _____

    c. _____   h. _____

    d. _____   i. _____

    e. _____   j. _____

28. Explain the ABC'S of South America. _____

    _____

    _____

    _____

NOTES

CHAPTER
6

EUROPE

AN
OVERVIEW

Europe, containing seven percent of the earth's land, is one of the smallest continents; yet, its people and culture have made an immense impact on the rest of the world. Through colonization and immigration, Europe exported not only its people, but its languages, religions, and cultures. Today, many people think of Europe as the great storehouse of the world's civilization and it is visited by millions of tourists each year.

Most tourists visit *Western Europe* as opposed to *Eastern Europe*. The division is based on centuries of cultural development as well as present-day political realities. The eastern frontiers of Western Europe are Finland, West Germany, Austria, and Italy. Those countries and everything west of them are Western Europe; everything east is Eastern Europe, except Greece, which is now considered a part of Western Europe, and Turkey, which is usually considered a part of the Middle East.

Western Europe amazes many North American tourists with the diversity of its cultures in such a small area. Although much of that diversity is a result of Europe's history, it also comes from its geography—its many peninsulas, islands, and the mountain ranges that divide it. The *Alps* are the most important mountain range of Western Europe. They reach their highest point at *Mont Blanc* in France, but are centered in Switzerland, the

The Pilatus, the Alps, Switzerland. *Photo: Courtesy Swiss National Tourist Office*

very heart or hub of Western Europe. It is the meeting place for three of the great European cultures—Italian, German, and French—and, indeed, all three languages are spoken there.

Let us use Switzerland as our hub of Europe and build around it, starting with the three cultures which surround it. South of Switzerland and separated from the rest of Europe by the Alps is the *Italian Peninsula* occupied by *Italy*. On Switzerland's north and east are the Germanic countries of *West Germany* and *Austria* and to the north and west is the country of *France*. Germany and France share a long border north of Switzerland until they are separated by three small countries: *Belgium, Netherlands,* and *Luxembourg.* These are often collectively called the *Benelux* countries.

We might very well consider these countries—Switzerland, Austria, Germany, France, and Benelux—the "core" of Western Europe because all the other areas are either peninsulas extending from it, like Italy, or are offshore islands. For example, the Iberian Peninsula or simply *Iberia,* occupied primarily by *Spain* and *Portugal,* extends west of France towards the Atlantic, Europe's western boundary. It is separated from the rest of Europe by the *Pyrenees Mountains.*

North of Germany are two major peninsulas. The smaller of the two is occupied by *Denmark,* the larger by *Finland, Sweden,* and *Norway.* The northern shores of the larger peninsula are bordered by the Arctic Ocean, Europe's northern frontier. These two peninsulas along with the island-nation of *Iceland* make up *Scandinavia.*

The most important islands off the mainland of Europe are the *British Isles,* occupied by the *Republic of Ireland* (*Eire*) and the *United Kingdom.* The United Kingdom, often called *Great Britain* or *Britain,* is actually composed of four countries united under the Crown: *England, Wales, Scotland,* and *Northern Ireland* (*Ulster*). The United Kingdom is separated from Scandinavia by the *North Sea* and from the continent itself by the narrow *English Channel.* The term, the *Continent,* meaning the landmass of Western Europe, was developed by the British who were—and are—very conscious of their separation from it.

The Mediterranean Sea, which forms Europe's southern border, is the home of many other European islands. The three largest are *Sicily* and *Sardinia,* which belong to Italy, and *Corsica,* which belongs to France. Spain's Mediterranean islands, called the *Balearic Islands,* include the very popular island of *Majorca.* The Greek islands, set in a part of the Mediterranean called the *Aegean,* are legendary for their beauty and ambience. While many people have their own special choices, *Mykonos* seems to be the favorite among most travelers.

Like European islands that share their identity with the sea surrounding them, some of Europe's cities share their identity with the famous rivers that flow by them. The best examples are London on the *Thames,* Paris on the *Seine,* and Vienna on the *Danube.* However, the major scenic waterway of the Continent is the *Rhine River.* The Rhine separates Germany and France in the south, flows north through German wine country, and empties into the North Sea at Rotterdam, one of the world's busiest ports. Another river beloved by tourists is the *Loire* in northwestern France for it flows through a beautiful valley dotted with magnificent châteaux.

Cahir Castle, Co. Tipperary, Ireland. *Photo: Courtesy Bord Fáilte Photo.*
Reproduction rights reserved.

No discussion of the waterways of Europe would be complete without mention of the *fjords* of Norway, those magnificent long and narrow arms of the sea which cut into the country's rugged coastline, or the beautiful alpine lakes of Switzerland. The *Sogne Fjord,* just north of Bergen, is considered Norway's most magnificent fjord. Among the lakes of Switzerland, there are the *Italian-Swiss Lakes,* naturally on the border with Italy, *Lake Geneva* on the border with France, and *Lake Constance* on the border with Germany and Austria.

Eastern Europe, dominated by the massive territory of the Soviet Union (USSR), also includes the *Balkan Peninsula,* extending south of Hungary between the *Black* and *Adriatic Seas.* Eastern Europe extends as far as the *Ural Mountains* and *Ural River* which separate it from Asia. The border between Europe and Asia continues from the Urals through the *Caspian Sea,* along the *Caucasus Mountains,* through the *Black Sea,* and then through the *Bosporus* straits which divide European and Asian Turkey.

Trajan Forum, Rome. *Photo: Courtesy Italian Government Travel Office*

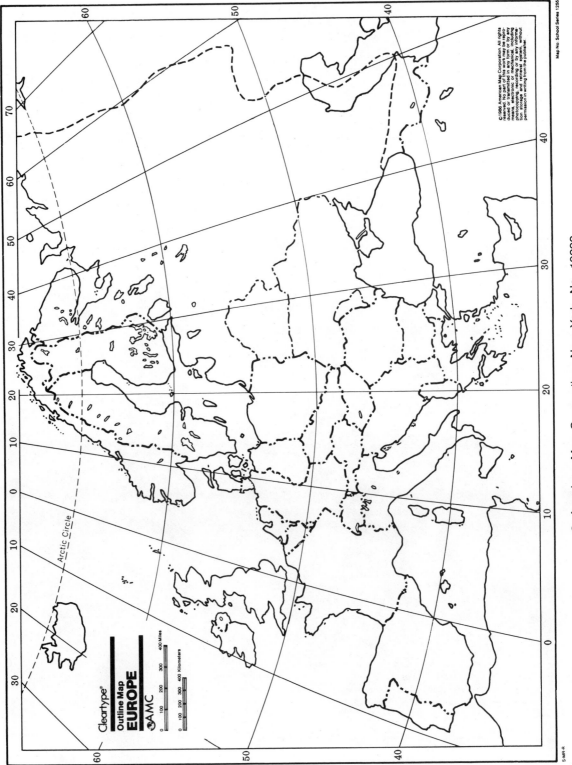

Cleartype®
Outline Map
EUROPE
AMC

0 100 200 300 400 Miles

0 100 200 300 400 Kilometers

Arctic Circle

Map No. School Series 1255

© American Map Corporation, New York, No. 19222

S-MR-R

EUROPE
An Overview

1. Name the eastern frontiers of Western Europe.

   _____

   _____

2. What mountains and river mark the eastern end of East-

   ern Europe? _____

   _____

3. Name the bodies of water which border Europe on its
   northern, western, and southern edges.

   _____

   _____

4. Which country do we consider the hub of Western Eu-

   rope? _____

5. Name the four main countries which border the hub of

   Western Europe. _____

   _____

6. Which countries make up Benelux? _____

   _____

7. In addition to Switzerland and Benelux, which other
   countries make up the ''core'' of Western Europe?

   _____

   _____

   _____

8. Name the countries of Scandinavia. _____

   _____

9. Which countries occupy the Iberian Peninsula?

_____

10. Name the two countries which share the British Isles.

_____

11. Which four countries make up the United Kingdom?

_____

_____

12. Name the dominant country of Eastern Europe.

_____

13. What are fjords and what country is famous for them?

_____

_____

14. What river provides a border between France and Germany? _____

15. What body of water separates the United Kingdom from France? _____

16. What mountain range separates Spain from the rest of Europe? _____

17. What mountain range separates Italy from much of the rest of Europe? _____

18. Name the peninsula occupied by Yugoslavia and its immediate neighbors. _____

19. To which country do each of the following islands belong?

a. Mykonos _____   c. Jersey _____

b. Corsica _____   d. the Orkneys _____

    e. Sicily _____   h. Majorca _____

    f. the Faeroes _____   i. Crete _____

    g. the Azores _____

20. Name the strait that separates Spain from the tip of North Africa. _____

21. Name the strait that divides European and Asian Turkey. _____

22. In what branch of the Mediterranean are the Greek islands located? _____

23. Name the two neighbor-countries of the mini-state of Andorra. _____

24. Name the country which surrounds the mini-states of San Marino and the Vatican. _____

25. Name the country which encircles Monaco on three sides.

_____

26. Liechtenstein is located between what two countries?

_____

27. Name the country in which each of these cities is located.

    Amsterdam _____   Edinburgh _____

    Athens _____   Florence _____

    Barcelona _____   Frankfurt _____

    Bergen _____   Geneva _____

    Brussels _____   Innsbruck _____

    Copenhagen _____ __   Istanbul _____

    Dublin _____   Lisbon _____

WHERE IN THE WORLD, WHEN IN THE WORLD?

Leningrad _____	Oslo _____
London _____	Paris _____
Madrid _____	Rome _____
Málaga _____	Salzburg _____
Milan _____	Stockholm _____
Monte Carlo _____	Venice _____
Moscow _____	Vienna _____
Munich _____	Zurich _____
Nice _____	

28. Place the following cities in order going from West to East: Barcelona, Lisbon, Madrid, Milan, Moscow, Munich, Nice, Vienna.

    a. _____   e. _____

    b. _____   f. _____

    c. _____   g. _____

    d. _____   h. _____

29. Place the following cities in order going from North to South: Berlin, Copenhagen, Florence, Munich, Oslo, Rome, Venice.

    a. _____   e. _____

    b. _____   f. _____

    c. _____   g. _____

    d. _____

30. The city of Berlin is in which two countries?

    _____

CHAPTER
7

EUROPE

THE UNITED KINGDOM
AND
THE CENTRAL CORRIDOR

Tourists who visit Europe from all over the world can find something of interest in virtually every corner of the continent. The countries of Europe are a treasure chest of quaint villages, exciting cities, magnificent cathedrals and castles, awe-inspiring museums, chic resorts, irresistible shops, ancient ruins, and spectacular scenery. To compile a compendium of these attractions would be well beyond the scope of this book and perhaps close to impossible at any rate. For our purposes, then, we would like to acquaint you with a selection of major European attractions, and, therefore, have emphasized the most frequently visited locations.

UNITED KINGDOM

Most North American visitors arrive at the gateway to the continent, *London, England.* The richness of its history, its wealth of cultural and artistic attractions, its sophisticated shopping areas, its splendid parks, gardens, and architecture draw tourists by the thousands.

London is almost synonymous with some of its key locations and institutions. *Westminster Abbey,* where the monarchs of the nation are crowned and in which many of the great persons in British life are enshrined, is a must on anyone's list of sites. Close by are the *Houses of Parliament* and *Big Ben,* the world's most famous clock and symbol of London itself. *No. 10 Downing Street,* home of the Prime Minister, rounds out the sites of government most tourists visit.

Buckingham Palace, home of the Queen, with its daily 11:30 a.m. ceremony of the *Changing of the Guard,* perhaps the world's most famous ceremony, provides visitors with a marvelous spectacle of the pageantry of Britain. The *Tower of London,* with its display of the crown jewels, is closely associated with the history of England. Among the many outstanding museums in London, the *British Museum* is preeminent and of the many magnificent churches, *St. Paul's Cathedral* is best known. *Hyde Park* with its famous speaker's corner ranks first among the many parks. *Harrod's* and *Fortnum and Mason* are the most famous of London's department stores.

The London theater is the most active in the world and concerts, ballets, and operas abound. Among the many performing centers in London are the *Theatre Royal* in Drury Lane and *Royal Festival Hall.* Many visitors delight in simply wandering through the streets of London stopping in such areas as *Picadilly Circus, Trafalgar Square,* and *Chelsea.* And, of course, no visit to London would be complete without a visit to an English pub.

Beyond London, visitors to the United Kingdom head for the two old and venerable universities, *Cambridge* and *Oxford,* the mysterious prehistoric ruins at *Stonehenge,* performances at the Shakespearean theater at *Stratford-upon-Avon,* and to *Edinburgh,* capital of *Scotland.* Rich in history and culture, Edinburgh offers visitors its castle, fortress of the Scottish kings; *Holyrood House,* home of Mary Queen of Scots; and *Princes Street,* which runs along well-tended parks with lovely views of the old city's skyline, often called the most beautiful street in Europe. In late summer, Edinburgh hosts an arts festival which draws performers and guests from all over the world.

Big Ben, London, England. *Photo: Courtesy British Tourist Authority*

Eiffel Tower, Paris, France. *Photo: Courtesy Air France*

Upon leaving the United Kingdom, most tourists travel south through France, Switzerland, and Italy; then turn north through Austria, Germany, and the Netherlands. This route is so popular that we might call it, along with London, the *Central Corridor* of Europe.

Traveling the Central Corridor allows tourists to enjoy most of the major attractions of Europe and a wonderful variety of scenery and cultures within a relatively short distance. For that reason, it is a favorite route of tour companies as well. Indeed, if you check the routings of the "grand tours" offered by many companies, you will find that they invariably follow all or part of the Central Corridor.

FRANCE

On our trip through the Central Corridor, we will go to *Paris* next. Rivaling London for wealth and breadth of attractions, Paris is famed for such streets and landmarks as the *Champs Elysées,* the *Arc de Triomphe,* and the *Eiffel Tower.* The *Louvre,* housing such works as the *Mona Lisa* and *Venus de Milo,* is one of the world's greatest museums; *Notre Dame* is one of the world's most beautiful cathedrals; and nearby *Versailles* is one of Europe's most spectacular palaces. Paris is, however, more than majestic landmarks. It is also sidewalk cafes, strolls along the *Seine,* beautiful parks and gardens, the artist's center known as the *Latin Quarter,* elegant shops, cozy bistros, and dazzling nightlife. No wonder it is often called the "world's most romantic city."

From Paris, visitors to France go to the châteaux country of the *Loire Valley,* the beaches of *Normandy* made famous in World War II, the legendary *Mont-Saint-Michel,* and of course the beach resorts of the *Riviera* along France's Mediterranean coast. While there, visitors often go to *Monte Carlo* in the mini-state of Monaco to gamble at or just to see the famed casino.

SWITZERLAND

Switzerland is synonymous with the Alps and spectacular scenery. It is also quaint villages and cities, sparkling lakes, watches, chocolates, and cuckoo clocks all wrapped up in what appears to be a make-believe land of charm, cleanliness, and efficiency.

Two of the higher mountains in the Swiss Alps are the *Jungfrau* (13,642 feet) and the *Matterhorn* (14,692 feet). Visitors may tour the mountains on guided excursions or view their magnificent scenery from the famous towns that surround them. The town of *Interlaken,* located near the Jungfrau, is a good base from which to hike, enjoy water sports on nearby lakes, or ride up the Jungfrau on Europe's highest railroad. One of the nicer stops on this tour is Jungfraujoch—at 11,333 feet, the highest point in Europe reached by rail. The town of *Zermatt,* near the Matterhorn, has the longest ski season in Europe. Also nearby is the world's most fashionable ski resort and celebrity playground, *St. Moritz.*

ITALY

Continuing south from Switzerland, we enter Italy. Visitors to Italy normally concentrate on the three major centers: Florence, Rome, and Venice. *Florence* is a living testament to the magnificence of the Renaissance. Some of the city's treasures are the *Piazza della Signoria,* a busy square and beautiful outdoor sculpture gallery; the cathedral and its accompanying bell tower, both magnificently decorated with colored marble; and the many

Klostus, the Alps, Switzerland. *Photo: Courtesy Swiss National Tourist Office*

works of Michelangelo, including his *David,* which is housed in the *Academia della Arte.* While every nook and cranny of this city seems to hold something of beauty, its highlights are on view in the *Uffizi Gallery,* one of the finest museums in the world.

Rome is, with London and Paris, part of the "Great Triad" of European cities. Rome, too, is synonymous with great streets and landmarks such as the glamorous *Via Veneto* where celebrities and film stars gather, the *Fountain of Trevi,* made famous in the movie *Three Coins in the Fountain,* and the *Spanish Steps*, adopted as an international meeting place particularly for the young. *The Sovereign State of Vatican City,* center of the Roman Catholic Church, with its *St. Peter's Cathedral,* the largest church in the world, is entirely enclosed by Rome. It draws many visitors from around the world.

Rome is rich in Renaissance art, architecture from many ages, and the ruins of ancient Rome. Michelangelo's *ceiling* of the *Sistine Chapel,* the magnificent *Piazza Navona* with its three fountains and surrounding churches, the *Colosseum,* and the *Forum* are sights all tourists must see. Modern Rome with its fine shopping, wonderful restaurants, and spirited cafe-nightlife also draws many visitors. Is it any wonder that Rome is called "the Eternal City"?

Turning north from Rome, we head for the romantic city of *Venice.* Known worldwide for its canals, gondolas, and red and white barber poles, Venice is truly one of the jewels of western civilization. Many of its famous palaces, churches, bridges, and other art and architectural treasures may be seen from its principal canal, the *Grand Canal,* one of the great waterways of the world.

The *Piazza San Marco,* called the "finest drawing room in Europe" by Napoleon, is the very heart of the city. Located just outside the *Cathedral of Saint Mark* and near the *Palace of the Doges* (the title of the city's earlier rulers), the Piazza draws many by its beauty and its lively outdoor cafe life.

It is difficult to leave Italy with still so much to see: the *Leaning Tower of Pisa,* the ruins of *Pompeii,* the *Isle of Capri,* but we also want to see Austria and the northern European countries. And so we understand the sweet conflict which many a tourist to Europe suffers!

AUSTRIA

Our first stop in Austria is *Innsbruck,* the country's center for alpine ski resorts and hiking trails. Most visitors also stop in *Salzburg*, famous to

Palace of the Doges, Venice, Italy. *Photo: Courtesy Italian Government Travel Office*

some from *The Sound of Music* and to others for its Mozart festival. It is *Vienna*, though, which draws most visitors to Austria.

Once the capital of the vast Austro-Hungarian Empire, Vienna is still an imperial city of cathedrals, palaces, parks, and cultural centers. Like the world's other great cities, Vienna has its famous street. Called the *Ring*, it follows the old city wall and includes many of the city's sights within its circle. Most famous of these are *St. Stephen's*, the city's great cathedral, and the *Hofburg*, palace of the royal Hapsburg family. At the Hofburg, visitors can enjoy two of Vienna's most distinctive attractions, the *Spanish Riding School*, with its well-trained, world-famous Lipizzaner stallions, and the *Vienna Boys' Choir*. The *Schonbrunn*, summer palace of the Hapsburgs and one of the most delightful palaces in Europe, is just outside the city as is its famous park, the *Vienna Woods*.

GERMANY

Continuing north into Germany, most tourists head for *Munich. Marienplatz*, the old heart of the city, draws many visitors. A favorite treat there is watching the mechanical figures of a clock, known as the *Glockenspiel*, perform daily at 11:00 a.m. For many visitors though, Munich means good beer and few leave the city without visiting the *Hofbrauhaus* or another of the famous beer halls where "ompahpah" music and conviviality are the order of the day. Beyond Munich, tourists visit the scenic area known as the *Black Forest* and fairy tale castles such as *Neuschwanstein*, so often depicted on travel posters.

Munich's major attraction, though, is its annual autumn festival, *Oktoberfest*, which attracts people from all over the world to its German beer, food, music, and entertainment.

Heading farther north, tourists follow the *Rhine River Valley*, sometimes taking a river cruise. They pass through the beautiful German wine country, stop to see the university at *Heidelberg* (made famous by the operetta *The Student Prince*) or the beautiful cathedral at *Cologne*, before going on to *Amsterdam* in the Netherlands (sometimes called Holland).

THE NETHERLANDS

Amsterdam is a graceful old city of canals and seventeenth-century homes. The *Anne Frank Home* and *Museum* draws many people as do the *Vincent van Gogh Museum* and the *Rijks Museum*, with the greatest collection of Dutch paintings in the world including many Rembrandts. Visitors to the Netherlands also enjoy going to diamond-cutting workshops, to cheese markets, and to the countryside to see the tulips and windmills which are universal symbols of this country.

Many North Americans complete their tour of Europe here and return home through London or Paris or even directly from Amsterdam itself, having enjoyed the variety of cultures and the many tourist attractions of the Central Corridor. Obviously, many tourists go beyond the Central Corridor either as extensions of the corridor or as separate trips.

The Netherlands

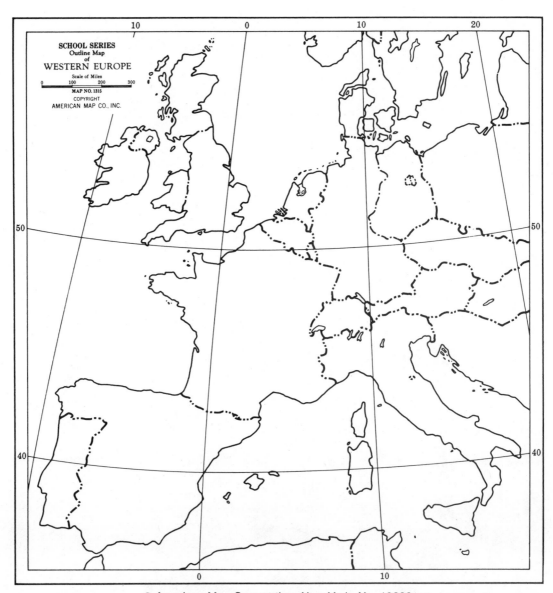

SCHOOL SERIES
Outline Map
of
WESTERN EUROPE
Scale of Miles
0 100 200 300
MAP NO. 1315
COPYRIGHT
AMERICAN MAP CO., INC.

EUROPE
The United Kingdom and the Central Corridor

1. Starting in England, list in order the countries visited on the Central Corridor route.

   _____

   _____

   _____

2. Name the three cities which make up the ''Great Triad'' of European cities.

   _____

3. Which city is generally regarded as the ''world's greatest city''? _____

4. Which city is called the ''most romantic city'' in the world? _____

5. Which city is known as the ''Eternal City''?

   _____

6. Big Ben is a symbol of which city?

   _____

7. The Eiffel Tower is a symbol of which city?

   _____

8. The Colosseum is a symbol of which city?

   _____

9. Where would visitors to the South of France go to gamble? _____

10. Who lives at No. 10 Downing Street?

   _____

11. Name the sovereign state completely enclosed within

    Rome. _____

12. Name the world-famous ceremony which takes place daily
    at Buckingham Palace at 11:30 a.m.

    _____

13. Name the most fashionable ski resort in Switzerland.

    _____

14. What is France's Mediterranean beach resort area called?

    _____

15. Where are the monarchs of Great Britain crowned?

    _____

16. Name the three major centers to visit in Italy.

    _____

17. Name the Swiss town which serves as a base for explor-

    ing the Jungfrau region. _____

18. In which city is each of the following streets or dis-
    tricts located?

    a. Princes Street   e. Picadilly Circus

       _____   _____

    b. Via Veneto _____ f. Spanish Steps

    c. Champs Elysées   _____

       _____   g. The Ring _____

    d. Latin Quarter

       _____

19. In which city is each of the following sites located:

a. British Museum

_____

b. Notre Dame

_____

c. Louvre _____

d. Forum _____

e. Holyrood House

_____

f. Uffizi Gallery

_____

g. Piazza Navona

_____

h. Hyde Park _____

i. St. Peter's

Cathedral _____

j. St. Paul's Cathedral

_____

k. Fountain of Trevi

_____

l. Piazza San Marco

_____

m. St. Stephen's

Cathedral _____

n. Spanish Riding

School _____

o. The Anne Frank
Home and Museum

_____

p. The Schonbrunn

_____

q. Glockenspiel

_____

20. Briefly identify and locate the following:

a. Cambridge _____

b. Loire Valley _____

c. Stonehenge _____

d. Matterhorn _____

21. Name the museum that has the greatest collection of Dutch paintings. Where is it located?

_____

22. Name Munich's world—famous annual celebration.

_____

23. The Student Prince is set in which university town?

_____

24. Which Austrian city is associated with The Sound of Music? _____

25. Name the town which is the center of the Austrian Alps.

_____

26. What is the name of the famous park on the outskirts of Vienna? _____

27. Name Venice's principal canal.

_____

28. On which river can tourists cruise through Germany's wine country? _____

CHAPTER
8
EUROPE

BEYOND
THE CENTRAL
CORRIDOR

IBERIA

Iberia is the most popular of the outlying areas with northern Europeans as well as North Americans. *Spain* and *Portugal* are a world unto themselves, separated from the rest of Europe by the *Pyrenees* and influenced by their historic close association with the Islamic cultures of North Africa.

While the beaches along the famed coasts, the *Costa del Sol* and *Costa Brava* in Spain and the *Algarve* in Portugal, are favored vacation spots, the cities of Iberia also attract may tourists with their history, art, entertainment, and ambience. *Madrid,* capital of Spain, is the major city in Iberia. It is famed for its museum, the *Prado*; bullfights, flamenco dancers, and distinctive foods such as paella and gazpacho. Beyond Madrid, visitors go to *Segovia* to see its beautiful castle (made famous in the film of *Camelot*) and its ancient Roman aqueduct; *Toledo,* hauntingly painted by El Greco; and *Granada,* with its unmatched *Alhambra Palace* distinguished by beautiful fountains, gardens, tiles, mosaics, and carvings.

GREECE, TURKEY, AND YUGOSLAVIA

Greece is, of course, a favored destination too. The *Acropolis,* center of the ancient city of *Athens,* is the center of sightseeing in the modern city. Crowned by the *Parthenon,* the ruins of the Acropolis are one of the greatest sites of Western civilization. After a day spent visiting the ruins, many visitors relax in the *Plaka,* known for its many taverns and nightlife. Beyond

Saint Sophia, Istanbul, Turkey. *Photo: Courtesy Turkey Office of the Culture and Information Attaché*

Athens, visitors head for other sites of antiquity, such as *Delphi,* famed for its oracle in ancient times, and, of course, the sun-drenched islands with their whitewashed villages. *Corfu* is the best-known island on the western side of the country, but *Mykonos* is the most popular of all the islands and attracts many, including artists and celebrities. *Crete* is the largest of all the Greek islands and, since it is farther south than all the others, it is very pleasant for winter vacations. *Rhodes* is a particularly beautiful island and has developed into a popular vacation center.

Visitors to Athens will often travel to *Istanbul,* Turkey as well. Istanbul, formerly *Constantinople,* has a long and rich history and is an especially interesting city because it has long been the bridge between Europe and the Middle East. Its bazaar is world famous as are its two best-known buildings, the *Blue Mosque,* a beautiful mosque dating from the seventeenth century, and *Santa Sophia,* a church of architectural wonder built in 532. Topkapi, the former residence of the sultans, pleases visitors with its many palaces, gardens, and beautiful views as well as its museum and its harem.

Many visitors return from Greece and Turkey through Yugoslavia. The favored destinations there are the Adriatic sea towns of *Dubrovnik* and *Split.* Both have preserved their old sections, and provide a graceful ambience, plenty of sunshine, and the sea.

Sogne Fjord, Norway. *Photo: Courtesy Norwegian National Tourist Office*

SCANDINAVIA

Scandinavia, north of the Central Corridor, is a totally different world from the Mediterranean worlds of Iberia, Greece, and Turkey. Scandinavia has more wilderness than any other part of Europe and so is a favorite destination for outdoor-oriented people. Most tourists head for *Copenhagen, Denmark* where they visit breweries, gaze at the *Little Mermaid,* a symbol of Copenhagen itself, and enjoy themselves at the *Tivoli Gardens,* the world-famous amusement park and garden.

They go on to *Stockholm, Sweden,* a picturesque city built on many islands. After visiting the old town, called *Gamla Stan,* visitors enjoy boat rides around the other islands of the city. The third Scandinavian capital, *Oslo, Norway,* is en route to the fjord country, considered by many to rival the Alps for scenic beauty in Europe. In Oslo, visitors see the remains of ancient viking ships and the famous *Kon Tiki* on display. *Bergen,* with spectacular views of its own, is the center for visiting the fjord country, either by ship or car or hiking. Of all the fjords, the *Sogne Fjord* is usually considered the most scenic.

SOVIET UNION

Many tourists combine a trip to the *Soviet Union* (USSR) with a trip to Scandinavia. *Leningrad* and *Moscow* are the two principal destinations in the USSR for Western tourists. Leningrad was known as *St. Petersburg* when, before the revolution, it was the czarist capital of Russia. Its imperial past is reflected in its beautiful buildings, parks, and tree-lined streets. It was built by Peter the Great, who admired the architecture of Western Europe and had his city built by French and Italian architects and craftsmen. Today, the former *Winter Palace* of the czars is a part of the *Hermitage Museum* which contains one of the world's greatest collections of art treasures, including more masterpieces of Western painting than any other museum. The most famous street in the city, *Nevsky Prospect,* draws many shoppers. Outside the city, tourists visit *Petrodvordets,* a preserved palace known for its beautiful grounds and glittering fountains.

Moscow is centered on *Red Square.* There, the city's two most famous sights, *Saint Basil's Cathedral* and the *Kremlin,* face each other. St. Basil's, built by Ivan the Terrible, is famous the world over for its many colorful onion-shaped domes. The Kremlin is, of course, the center of power in the USSR. It is a fortress which contains not only government offices but also an excellent museum and many beautifully preserved cathedrals from the czarist era. The *Mausoleum,* where Lenin's embalmed body is on display, is just outside the Kremlin. For lighter fare, many visitors enjoy an excursion to *GUM,* Moscow's famous department store.

Some tourists continue their journey through the USSR on the *Trans-Siberian Express,* one of the great train trips of the world. Traveling almost 6,000 miles over the world's longest continuous track, the train takes eight days and passes through seven time zones to go from Moscow to *Nakhodka,* a port on the Pacific coast of Asia just outside *Vladivostok.* A great adventure, indeed, which leads us to our own visit of Asia in the next two chapters.

Piran, Istra Peninsula, Yugoslavia. *Photo: Courtesy Yugoslav National Tourist Office*

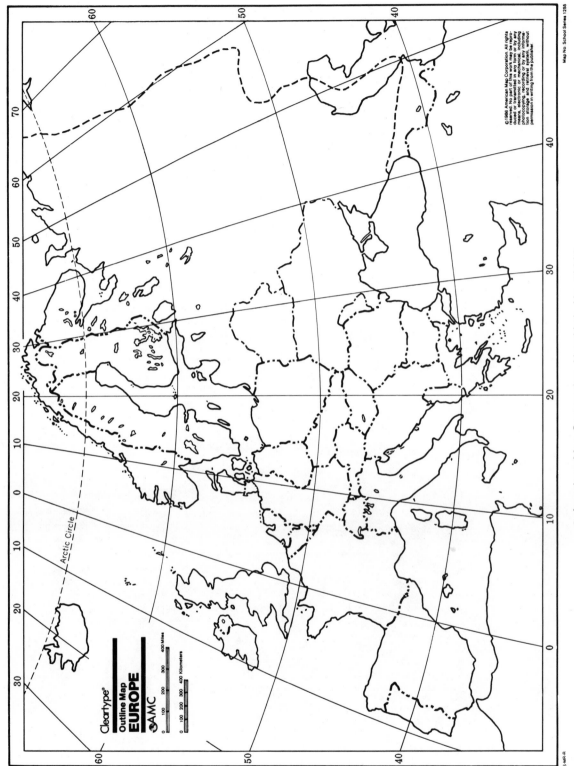

Cleartype®
Outline Map
EUROPE
AMC

Arctic Circle

Map No. School Series 1255

© American Map Corporation, New York, No. 19222

S-MR-R

EUROPE

Beyond the Central Corridor

1. Which city is the gateway to the Norwegian fjord country? _____

2. Where is the castle featured in the film <u>Camelot</u> located? _____

3. Name the most popular of the Greek islands.

   _____

4. Name the heart of Stockholm's old section.

   _____

5. Name the most scenic of Norway's fjords.

   _____

6. Name Copenhagen's world-famous amusement park.

   _____

7. Name and locate by country three popular beaches in Iberia.

   _____

   _____

   _____

8. Which museum has more masterpieces of Western painting than any other? Where is it located?

   _____

9. The Trans-Siberian express runs from Moscow to which city? _____

10. Name two Yugoslavian tourist centers on the Adriatic Sea. _____

11. In which city would you find each of the following:

    a. The Little

       Mermaid _____

    b. Kon Tiki _____

    c. The Kremlin _____

    d. The Alahambra

       Palace _____

    e. The Blue Mosque

       _____

12. Name the colorful cathedral on Moscow's Red Square.

_____

13. What is the name of Madrid's famous museum?

_____

14. What attraction is the center of sightseeing activities in Athens? _____

15. Name Leningrad's best-known street.

_____

16. Name the largest of the Greek islands. _____

17. Where is the tavern and entertainment section of Athens? _____

CHAPTER
9
ASIA

The Mideast
and
The Indian Subcontinent

Asia is by far the largest and most populous continent of all. It contains thirty percent of the world's land and sixty percent of the world's people. It stretches from the Arctic wastelands of Siberia to the steaming jungles of the equator to the endless deserts of Arabia. Its shores are washed by three of the world's four oceans, the Arctic, the Indian, and the Pacific, as well as the Mediterranean Sea. It contains both the world's highest point, *Mount Everest,* and lowest point, the *Dead Sea.* Its vast peninsulas and islands are worlds of their own.

Asia is connected to Africa by a thin land bridge and to Australia by a string of islands. Its "connection" to Europe is so extensive that some geographers consider Europe one of the peninsulas of Asia. Generally, though, the border between the two continents is drawn along the *Ural Mountains,* the *Ural River,* the *Caspian Sea,* the *Caucasus Mountains,* the *Black Sea,* and the *Bosporus,* the straits which separate European and Asian Turkey.

Because Asia is so vast and varied in climate, landscape, and culture, geographers have subdivided it to deal with it more easily. As is the case with many geographic divisions, the borders are fuzzy, but here you only need a general sense of those divisions. Let us start at the top of the world, the *tableland of Tibet,* averaging 16,000 feet, and backed by the world's greatest mountain range, the *Himalayas.* This tableland may well be considered the heart or the hub of Asia with the various subdivisions radiating out from it like spokes in a wheel.

The *Far East* is probably the best-known subdivision of Asia. The term was developed by the British to refer to those areas that were the farthest east of Britain. Today, the term refers to those countries, east of the tableland of Tibet, at the eastern side of Asia: *People's Republic of China, Mongolia,* the *Korean Peninsula* occupied by *North* and *South Korea,* the island-nation of *Japan,* and the British Crown Colony island, *Hong Kong.*

Since the days of the Vietnam War, the area south of the Far East, as described above, has generally been called *Southeast Asia.* It includes the *Indochina Peninsula*—logically enough the peninsula between India and China. *Malaysia, Burma, Thailand, Laos, Kampuchea* (*Cambodia*) and *Vietnam* are located here as are the island-countries of *Singapore, Indonesia,* and the *Philippines.*

The major peninsula in Asia is occupied by *India.* This peninsula, along with the countries which neighbor India, is called the *Indian Subcontinent* because it covers such a vast area. Simply put, it is the area south of the tableland of Tibet, separated from it and the rest of Asia by the Himalayas. In addition to India it includes *Pakistan, Bangladesh,* the island-country of *Sri Lanka* (formerly *Ceylon*), *Nepal,* and *Bhutan.*

The rest of Asia, south of the Soviet Union, from Afghanistan to the Mediterranean, used to be divided between the Near East and the Middle East, but today the activities of the region are so intertwined that the entire area is collectively called the *Middle East* or more simply the *Mideast.* This is a loose term and has even been used to refer to Egypt and parts of North Africa. Our use of the term includes *Afghanistan, Iran, Iraq, Turkey,*[1] *Syria, Jordan, Lebanon, Israel,* and the countries of the *Arabian Peninsula.*

---

[1]Turkey was discussed with Europe in Chapter 8 since it is most often visited along with Greece and Yugoslavia.

Western Wall, Jerusalem, Israel. *Photo: Courtesy Israel Government Tourist Office*

North of all the subdivisions we have discussed and stretching from the Ural Mountains to the Pacific Ocean lies the Asian territory of the Soviet Union, nearly forty percent of the continent. The Soviet territory falls into two subdivisions. *Soviet Central Asia* is a rich and historic land north of Afghanistan, between western China and the Caspian Sea. North of Soviet Central Asia, *Siberia* extends across the breadth of the continent. One of the most remote areas of the world, it is rich in minerals and natural resources. It is the eastern tip of Siberia which is only a few miles away from Alaska across the *Bering Strait.*

To review then, the vast continent of Asia is subdivided into six areas for easier study. These include the *Far East, Southeast Asia,* the *Indian Subcontinent,* the *Mideast, Soviet Central Asia,* and *Siberia.* The obvious missing term is the *Orient,* which, though not a geographic term, is still commonly used to refer to those parts of Asia that we have placed in the *Far East* and *Southeast Asia.*

THE MIDEAST

Israel is the first Asian experience for many North American tourists, since this country is often included in tours of Europe. Furthermore, religious pilgrims flock there from all over the world to visit sites and shrines sacred to Christians, Jews, and Moslems.

Although *Tel Aviv* is the largest city in Israel, *Jerusalem* is of primary interest to tourists. There, amidst the intriguing Old City, visitors encounter many religious shrines and biblical sites including the *Western Wall,* the holiest place in the world to Jews, and the *Garden* of *Gethsemane,* so important to Christians. Visitors also go to the *Biblical and Archaeological Museum* where they can see the *Dead Sea Scrolls.*

To the south of Jerusalem lies *Bethlehem,* birthplace of Christ and destination of many pilgrims at Christmas time. The *Dead Sea,* the lowest point on earth at about 1,300 feet below sea level, is also to the south of Jerusalem. This area is dominated by *Masada,* a mesa-shaped rock which was the scene of the heroic resistance of the Jews to Roman rule. It has gained wide fame through a novel and a television series on the subject.

We should not leave the *Mideast* before mentioning *Mecca* in *Saudi Arabia.* Although this city is closed for religious reasons to most North Americans, it is the holiest city for Moslems and receives thousands of religious tourists or pilgrims every year.

THE INDIAN SUBCONTINENT

Turning to the *Indian Subcontinent,* we should note that *Calcutta* is India's chief port, commercial center, and largest city and that *Bombay,* her second largest city, is her most cosmopolitan and is usually regarded as her gateway. It is to the north, however, to *New Delhi,* that most tourists go.

New Delhi, capital of India, is built upon centuries of other cities and combines the country's ancient and modern characteristics. Ancient temples, palaces, forts, and modern government sites as well as its vast bazaar make the city an interesting one to visit.

The most famous of India's sights and, indeed, one of the most famous in the world, is the *Taj Mahal* in *Agra.* This magnificent marble mausoleum is considered one of the most beautiful buildings in the world. Some even say that to see it on a moonlit evening is worth the trip to India. Tourists will also go to the nearby cities of Jaipur and Varanasi (Banares). Medieval *Jaipur* is famous as the "Pink City" because of the lovely color of its buildings. *Varanasi,* on the banks of the holy *Ganges River,* is considered the premier city of Hinduism. Here, millions of pilgrims come to wash their sins away in the sacred river.

North of New Delhi lies *Kashmir,* one of the world's most beautiful areas with lakes, rivers, mountains, and meadows. *Srinagar,* the capital of Kashmir, lies along *Lake Dal* where many tourists stay on comfortable, if not deluxe, houseboats.

Nepal, India's neighbor, also attracts many tourists. They are drawn there by the very remoteness and spectacular scenery of the *Himalayas,* the world's highest mountains. *Katmandu,* capital of the kingdom and quite

Festival Dussera parade, Temple Tower in background, Mysore, India.
Photo: Courtesy Government of India Tourist Office

isolated until very recently, has a medieval flavor which intrigues visitors from the "modern world." Hardy tourists set out for treks in the mountains while others are content with sightseeing by plane. In any case, all hope for a clear view of *Mount Everest,* at over 29,000 feet, the world's highest mountain.

In our next chapter, we will look at the attractions of the *Far East* and *Southeast Asia,* together called *The Orient.*

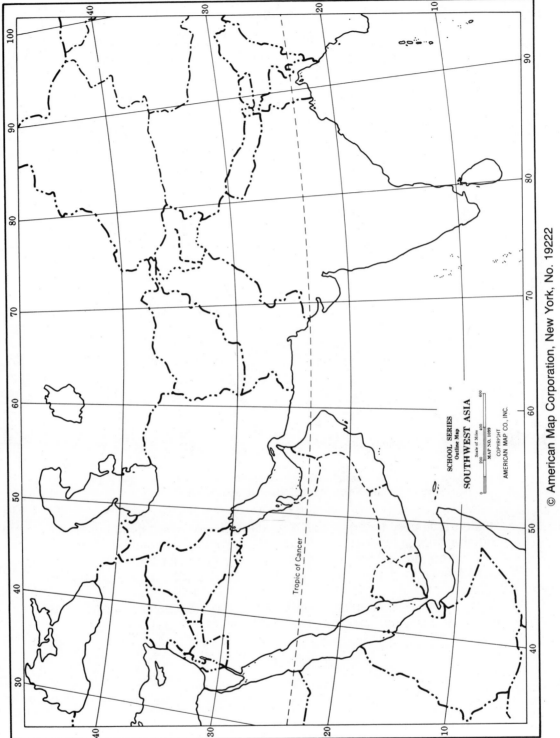

SCHOOL SERIES
Outline Map
SOUTHWEST ASIA

Scale of Miles
200   400   600

MAP NO. 1099
COPYRIGHT
AMERICAN MAP CO., INC.

Tropic of Cancer

ASIA
The Mideast and the Indian Subcontinent

1. Asia contains what percent of the earth's land? Of the

   earth's people? _____

2. Name the three oceans which border Asia.

   _____

3. Name the six geographic features which provide the
   border between Europe and Asia.

   _____

   _____

   _____

4. What geographic feature may be considered the heart of

   Asia? _____

5. Name the six subdivisions of Asia.

   _____

   _____

   _____

6. In which subdivision would you place each of the following peninsulas:

   a. Korean _____   c. Indochina _____

   b. Arabian _____   d. Indian _____

7. In which subdivision would you place each of the following islands or groups of islands?

   a. Hong Kong _____   c. Sri Lanka
                                   (Ceylon) _____

   b. Japan _____   d. Indonesia _____

8. Place each city below in its proper country. Then place each country in its proper subdivision.

City	Country	Subdivision
a. Peking	_____	_____
b. Tokyo	_____	_____
c. Manila	_____	_____
d. Bangkok	_____	_____
e. Tashkent	_____	_____
f. Tel Aviv	_____	_____
g. Karachi	_____	_____
h. Vladivostok	_____	_____
i. Bombay	_____	_____
j. Amman	_____	_____
k. Seoul	_____	_____

9. What and where is the highest point in the world?

_____

10. What and where is the lowest point in the world?

_____

11. Name the highest mountain chain in the world.

_____

12. What is the major city in Israel for tourists?

    _____

13. Name and locate the holiest city in the world for Mos-

    lems. _____

14. Name and locate the most important city in the world for

    Hindus. _____

15. Name India's beautiful mountain region, the region's
    capital, and its major lake.

    _____

    _____

16. Name the ''Pink City'' of India. _____

17. Name the holy river of the Hindus. _____

18. What and where is India's most famous tourist attrac-

    tion? _____

19. Name India's capital. _____

20. What percent of Asia's landmass is in the Soviet Union?

    _____

21. Name the two subdivisions of Asia in Soviet territory.

    _____

22. What body of water separates Siberia from Alaska?

    _____

23. What body of water separates Japan from Mainland Asia?

    _____

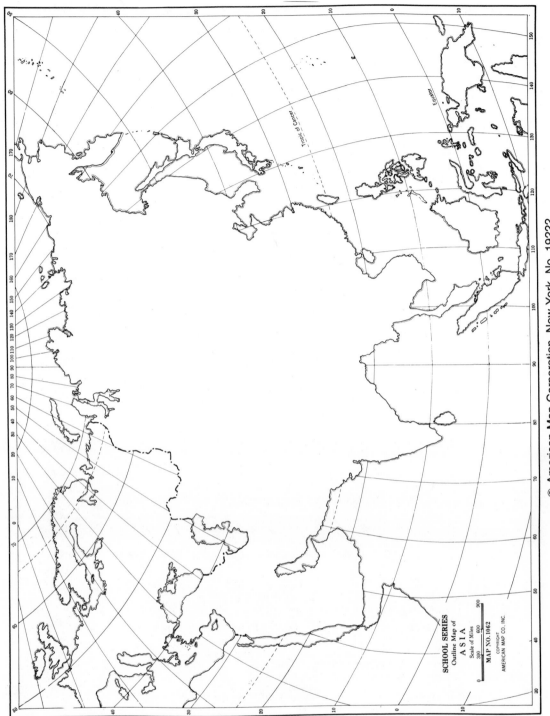

SCHOOL SERIES
Outline Map of
A S I A
Scale of Miles

MAP NO. 1062
COPYRIGHT
AMERICAN MAP CO. INC.

Tropic of Cancer

Equator

© American Map Corporation, New York, No. 19222

24. Name the countries that border Soviet Asia.

_____

_____

25. Name the countries that border Israel.

_____

_____

NOTES

CHAPTER
10

ASIA

The Orient

The term the Orient rings with the colorful, the exotic, and the mysterious in the minds of many people. The Orient, which literally means the East, is made up of the Far East and Southeast Asia. Its magic draws many people and makes it the major destination for tourists to Asia.

Most visitors travel there on tours. While tours will differ from company to company, many do take their clients to the same points of interest. The most common itinerary for a tour to the Orient includes all or some of the following: Tokyo and other points in Japan, Taipei in Taiwan, Hong Kong, Bangkok in Thailand, Singapore, and Bali in Indonesia. Hong Kong and Japan are also popular as single-destination trips. The *People's Republic of China*—hereafter simply *China*—is usually a single-destination trip, but is sometimes combined with visits to Hong Kong and/or Japan.

JAPAN

Japan combines all the vitality of the modern industrial age with the tranquility of formal gardens and ancient tea ceremonies. It draws, therefore, both business travelers and tourists. *Tokyo,* the country's capital, is the gateway to Japan and to all Asia. It is one of the world's largest and busiest cities and is the busiest air transportation center in Asia. Tokyo's bustling modern side is best known to tourists from the *Ginza,* its world-famous shopping area. In contrast, the *Emperor's Palace* is set in 250 acres of landscaped gardens in the heart of the city. Other attractions in Tokyo include the *Diet Building,* home of the Parliament, many museums, and cultural facilities. Adjacent to Tokyo is *Yokohama,* its port city.

Osaka is Japan's second major city. It is an important commercial center. Its primary tourist attraction is the *Osaka Castle,* built in the sixteenth century and especially famous today for its role in the popular novel and television special *Shogun.* The country's ancient capital, *Kyoto,* often called the jewel of Japan, is comparable to Florence in western civilization. Kyoto is a treasure house of art, architecture, temples, shrines, palaces, and landscaped gardens. Indeed, it is said that this city embodies the very spirit of Japan.

Another ancient city, *Nara,* also contains many examples of ancient architecture and is the home of the country's largest Buddha. Set in the *Todaiji Temple,* this eighth century Buddha towers fifty-three feet high. *Hiroshima,* with its memorials to the atomic bomb blast of 1945, is a stark contrast to the ancient splendors of Kyoto and Nara.

Cities are, however, not Japan's only attraction. The country is also justly famous for its natural beauty, especially *Mount Fuji.* This majestic 12,000-foot peak, covered with snow most of the year and surrounded by a semicircle of five lakes, is the very symbol of Japan. Mt. Fuji is in *Izu-Hakone-Fuji National Park,* a region of spas, hot springs, and many resorts. *Nikko National Park* is well known for its mountains, lakes, and waterfalls and is regarded as the most beautiful park in Japan. The *Inland Sea* is another very popular scenic area. This body of water separates the islands of Honshu and Shikoku. Many visitors enjoy a cruise among its 600 islands.

Public transportation and tourist accommodations are excellent throughout Japan. Tourists especially like riding on the 130-mph *Bullet Train,* until recently the world's fastest train and still a model of advanced rail technology. In contrast to that very modern experience, many tourists

Mount Fuji, Japan. *Photo: Courtesy Japan Air Lines.*

choose to stay in the traditional inns of Japan, called *Ryokans.* There, guests don traditional kimonos and learn to enjoy the Japanese baths. While visitors go to Japan for business and pleasure all year, the spring and autumn are especially popular, the first for the highlights of the gardens, the second for the warm, sunny weather and mountain foliage.

CHINA

While Japan draws more visitors than any other country in Asia, the most exciting adventure for travelers to Asia is a visit to *China.* Given the differences between its culture and that of the Western world and the isolation it maintained until recently, China has a very special appeal for North Americans and Europeans.

Ranking third in size among all nations, China is home to over twenty percent of the earth's people. More important than size and number, however, China has the world's oldest continuous civilization. China's culture has influenced that of many of her neighbors, including powerful Japan. Her civilization, therefore, may be considered one of the most important and influential in the world.

Nearly all tourists to China visit *Beijing* (Peking), the capital of the country. In this city, travelers confront both China's ancient imperial heritage and its modern revolutionary legacy. The *Forbidden City,* which contains the *Imperial Palace* and other buildings associated with the Chinese monarchy, is in the very center of Beijing. Nearby is *Tien Ah Men Square,* the immense public square which has played such an important part in China's revolutionary history. There, visitors will see the *Great Hall of the People* and the *Mao Tse-tung Mausoleum.*

Outside Beijing, the *Great Wall of China* is the primary tourist attraction. This 3,600 mile wall is one of the great achievements of mankind and is the only man-made object visible from outer space. It is unquestionably one of the world's great sites.

To date, there is no set itinerary of China; it varies from tour company to tour company. Certain points of interest seem to get more attention than others, however. Cruises down the *Yangtze River,* China's great river, are a popular feature on many tours as are visits to the cities of Guilin, Shanghai, Suzhou, and Xi'an.

Guilin is the center of a very scenic area where limestone mountains rise straight up over 1,000 feet to the sky. Over the centuries, erosion has created a fantastic landscape complete with caves and underground channels.

Shanghai, China's largest and most modern city, is one of the most populous cities in the world. It is a major commercial center and provides an interesting contrast to traditional China. *Suzhou* is a network of canals and ponds and is well known for its gardens, said to be the most beautiful in China.

Xi'an is noted for the archaeological wonders which have been excavated recently in its area. The artifacts on display give a wonderful insight into China's ancient past.

HONG KONG

Hong Kong, though, may be the best-known Chinese city. Still a British colony, it combines East and West. Moreover, it combines the hustle and bustle of one of the world's modern financial and banking centers with the romance of the recent colonial past.

Its harbor is one of the world's most beautiful; its shops and restaurants are legendary. As with other beautiful cities in the world such as Rio de Janeiro and Cape Town, Hong Kong has a tram to a high point, known as the *Peak,* from which the visitor gains a magnificent view of the city, the harbor, and the outlying islands.

Hong Kong's role as the emporium of the world is being challenged by *Singapore,* another small island-country with considerable Chinese cultural influence, which has become one of the world's great commercial, trading, and shopping centers.

Beyond Japan, China, and Hong Kong, *Bangkok* in Thailand and *Bali* in Indonesia are very popular points on tours of the Orient. *Bangkok* is especially well known for its colorful floating market and lovely temples, and *Bali* is a bit of paradise on earth with its beautiful beaches, resorts, temples, and festivals. So beautiful is Bali that it often is called the island of the gods.

As difficult as it is to leave anyplace on such a note, our next chapter draws us to the mystery and beauty of *Africa.*

Hong Kong

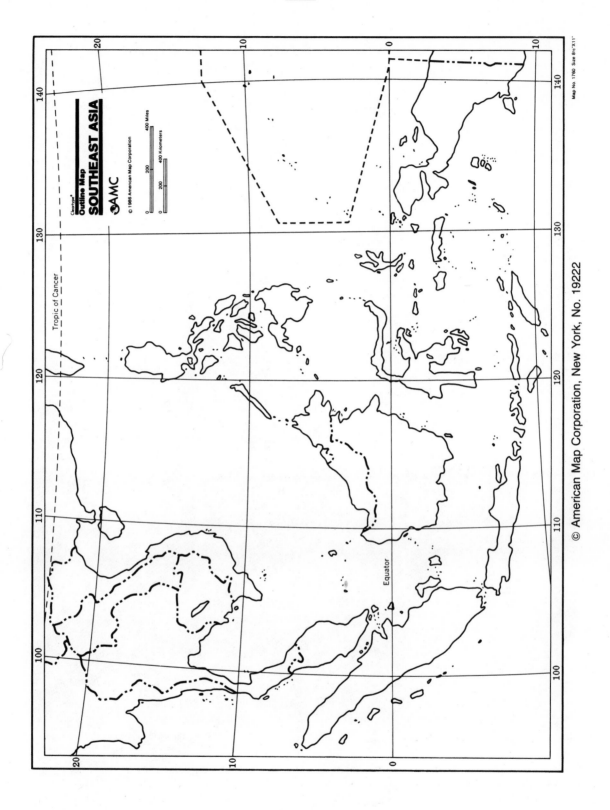

Cleartype®
Outline Map
SOUTHEAST ASIA

AMC

© 1985 American Map Corporation

400 Miles

400 Kilometers

Tropic of Cancer

Equator

© American Map Corporation, New York, No. 19222

Map No. 1760 Size 8½"X11"

ASIA
The Orient

1. Which two subdivisions of Asia are included in the term, the Orient? _____

2. What does the Orient literally mean?

   _____   _____

3. Which city is the busiest air transportation center in Asia? _____

4. Which city embodies the ''spirit of Japan''?

   _____

5. Name the natural feature which is the very symbol of Japan. _____

6. Name Japan's exceptional train.

   _____

7. What are traditional Japanese inns called?

   _____

8. In which country is the world's oldest continuous civilization? _____

9. Where does China rank in size among nations of the world? _____

10. What percent of the world's population lives in China?

    _____

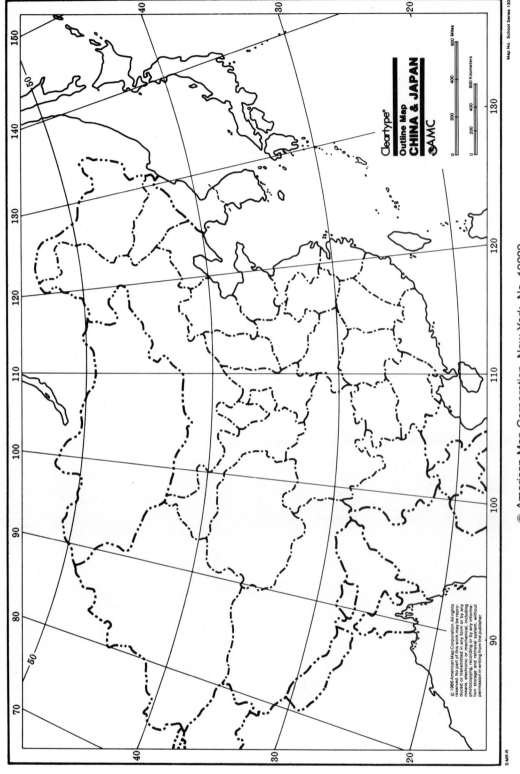

11. Name China's capital. _____

12. Name China's largest and most modern city.

    _____

13. On which Chinese river are cruises for tourists avail-

    able? _____

14. Which city could be considered the garden city of China?

    _____

15. Name the British Crown Colony in the Orient.

    _____

16. Name the city known for its floating market.

    _____

17. Name the Indonesian resort island famed for its beauty.

    _____

18. Name the area of Beijing which contains the Imperial

    Palace. _____

19. Name the great public square of Beijing.

    _____

20. Name the world—famous shopping area of Tokyo.

    _____

21. Which Chinese city is at the center of important ar-

    chaeological finds? _____

22. What is the only man—made structure visible from outer

    space? _____

23. Starting in Beijing and moving in a clockwise direction, place the following cities in order: Bangkok, Jakarta, Kuala Lumpur, Manila, Rangoon, Seoul, Singapore, Taipei.

    a. _____   f. _____

    b. _____   g. _____

    c. _____   h. _____

    d. _____   i. _____

    e. _____

24. Name the city which rivals Hong Kong as the emporium of

    the world. _____

25. Which mountain park is regarded as Japan's most beau-

    tiful? _____

CHAPTER
11
AFRICA

Known as the Dark Continent because of its late exploration, Africa still casts a spell of the mysterious and unknown over the world's imagination. Our images of Africa range from the barren dunes of the Sahara Desert to the impenetrable jungles of the Congo and from crowded, colorful bazaars surrounded by minarets and mosques to wide-open rolling grasslands sprinkled with herds of zebra and giraffe. These contrasting images actually define the two separate worlds of Africa, Africa north of the Sahara and Africa south of the Sahara.

NORTH AFRICA

Africa north of the Sahara or simply North Africa is truly a world apart from the rest of Africa. The region can easily be defined as those countries across the top of Africa which border the Mediterranean Sea: Morocco, Algeria, Tunisia, Libya, and Egypt. These countries are separated from the rest of Africa by the world's greatest desert, the *Sahara,* a barrier as formidable as any ocean or mountain range.

The historical and cultural distinctions of North Africa, however, have been as powerful as the desert in separating it from the rest of Africa. Unlike other areas of Africa, North Africa was a part of the ancient Mediterranean world with close ties to the Phoenicians, the Greeks, and the Romans. Today, its people are Arabic; its culture is Islamic; and vast deposits of oil have been found in some of its countries. Furthermore, it is physically connected to the Middle East by the *Sinai Peninsula.* The Sinai is traversed by the *Suez Canal,* so vital to the world's navigation and flow of oil. These factors, along with the physical connection via the Sinai Peninsula, more closely link the region with the Middle East than with the rest of Africa today.

Although both Morocco and Tunisia enjoy a brisk tourist trade, *Egypt* is the foremost tourist destination in North Africa. Nourished by the *Nile,* the longest river system in the world, Egypt developed one of the world's great ancient civilizations. Today, the surviving monuments of that great ancient civilization nourish Egypt's travel industry, drawing visitors from everywhere.

Cairo is the heart of Egypt and one of the world's major cities. It is Africa's largest city and the center of the Islamic world. Tourists in Cairo view the world-famous *Pyramids* and the *Sphinx,* gaze at the fantastic collection of artifacts from the ancient Egyptians, including mummies and the world-famous King Tut treasures in the *Egyptian Museum,* and bargain at the *Khan El Khalili bazaar.* From Cairo, many tourists take boat cruises up the Nile to the city of *Luxor,* site of the awesome ruins of the *Temples of Karnak* and *Luxor* as well as the *Valley of the Kings* and the *Valley of the Queens.* The valleys are the burial areas of the ancient royalty. There, many tombs are opened so visitors can enter them, see the murals, understand the construction, and taste the mystery of ancient Egypt.

Farther up the Nile, tourists will visit the *Aswan Dam,* and take the short flight to *Abu Simbel,* regarded by many as the most splendid of the Egyptian monuments. This site with its famous four statues of Ramses is not only a superb testimonial to the achievements of this ancient civilization,

Temple of Karnak at Luxor, Egypt. *Photo Courtesy Ministry of Tourism, Cario, Egypt*

but to those of modern civilization as well, for through modern technology, the complete site was removed from its original location, to avoid the floods caused by the dam's lake, and reconstructed on a new location.

Below Egypt and the other countries of North Africa, there is a tier of *transition countries* across the Sahara. We call them transition countries because of their ethnic and cultural mix, primarily Arabic in the north, and predominantly black in the south.

AFRICA SOUTH OF THE SAHARA

Below the Sahara Africa may be divided into West, Central, East, and Southern Africa. These divisions are based on geography, climate, history, economics, and people. *West Africa* is commonly used to group those smaller countries from Nigeria west to the coast below the Sahara on the great hump of Africa. *Nigeria,* with its great oil deposits, is the best known of these countries. *Central Africa* includes *Central African Republic* (naturally enough), *Zaire,* and several surrounding countries. Neither West nor Central Africa is of primary importance to the tourist industry.

On the other hand, *East Africa* contains the jewels of tourism in Africa including *Kilimanjaro,* Africa's highest mountain, and the *Serengeti Plains,* one of the greatest wildlife areas in the world. Both are located in

Tanzania. Lake Victoria, Africa's largest lake, borders Uganda, Kenya, and Tanzania, and is a favorite tourist stop. Kenya is also famous for its extensive game reserves, national parks, and beach resorts.

In fact, *Kenya* is the area's major tourist-attracting country with its world-famous parks, *Tsavo, Amboseli, Samburo,* and *Masai Mara,* where visitors can see the entire range of big African game in their natural environments. Among the many lodges which serve tourists in these regions, *Treetops Hotel* is the most famous. Although Kenya's animals provide its primary appeal to tourists, the country has also developed extensive resorts along its Indian Ocean beaches, particularly around *Mombasa.* Its capital city, *Nairobi,* is the major city in East Africa and is the commercial and transportation center of the region.

SOUTHERN AFRICA

Southern Africa stretches south of Tanzania and Zaire. We use Southern rather than South to distinguish the region from its major country, *South Africa.* Southern Africa is distinguished from the rest of Africa by the extensive settlement and influence of modern European people, such as the Portuguese, Dutch, and British.

Table Mountain, Cape Town, South Africa. *Photo: Courtesy SATOUR*

The greatest spectacle in Southern Africa is *Victoria Falls* on the *Zambezi River* between *Zimbabwe* and *Zambia*—the best viewing is from the Zimbabwe side. Otherwise, most tourists to this region spend most of their time in the country of *South Africa.*

South Africa's major city is *Johannesburg* which serves as the commercial and transportation center of the region. Tourists can visit the gold and diamond mines near Johannesburg, which help to make this country so wealthy. Most people, however, head for *Kruger National Park,* the country's major wild animal park; *Durban,* center of the beach resorts on the Indian Ocean; and especially historic *Cape Town,* the first European settlement in the country.

Situated on a harbor near the base of *Table Mountain* at the *Cape of Good Hope,* Cape Town has one of the world's most beautiful settings. The top of Table Mountain can be reached by a 4,000-foot cable-car journey for spectacular views of the city and the Cape area. The Cape area is rich in both history and natural beauty. It is especially well known for its distinctive Cape Dutch architecture and bountiful vineyards. These two features come together most gracefully at *Groot Constantia,* the oldest vineyard in the country, built in the late 1600s, and at *Stellenbosch,* a beautifully preserved historic town deep in the wine country.

Cape Town is connected to Port Elizabeth on the East Coast by the scenic *Garden Route* highway. In addition to providing beautiful views, this highway leads to beaches, resorts, and ostrich farms where visitors can actually ride the giant birds. Cape Town is connected to Johannesburg and beyond to *Pretoria* by the fabulous *Blue Train,* considered a five-star hotel on wheels and widely regarded as the world's most luxurious train.

Now that we are at the southern tip of Africa, we are ready to sail off to *Oceania.* And sailing would be a wonderful way to visit this continent for it is truly an ocean dotted with islands, among them, the giant Australia.

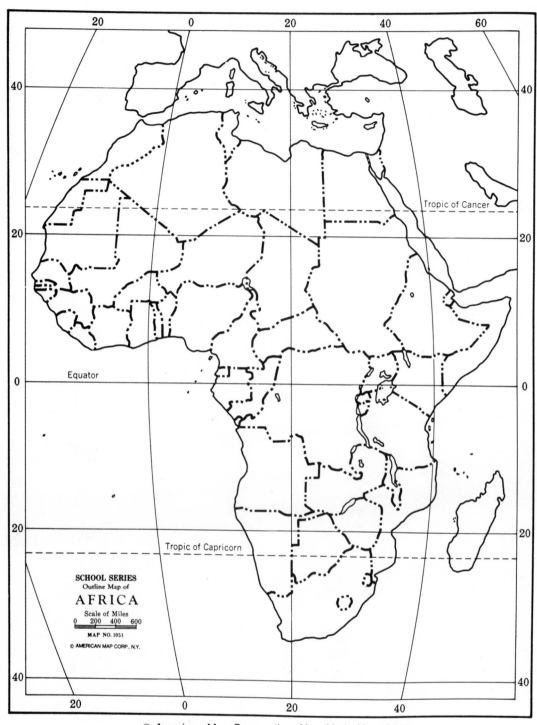

SCHOOL SERIES
Outline Map of
AFRICA
Scale of Miles
0   200   400   600
MAP NO. 1051
© AMERICAN MAP CORP., N.Y.

Tropic of Cancer

Equator

Tropic of Capricorn

AFRICA

1. What are the two separate worlds of Africa?

   _____

2. Briefly define North Africa.

   _____

   _____

   _____

3. Name the world's greatest desert.

   _____

4. Which country borders both the Mediterranean Sea and the
   Atlantic Ocean? _____

5. Which two bodies of water does the Suez Canal connect?

   _____

   _____

6. Which country borders both the Indian and Atlantic
   Oceans? _____

7. Name the famous cape near the southern tip of Africa.

   _____

8. Name the largest island off the coast of Africa.

   _____

9. Name the countries which border Kenya.

   _____

   _____

   _____

10. Name the largest city in Africa.

_____

11. Name the major city of East Africa.

_____

12. Name the major city in Southern Africa.

_____

13. Where was South Africa's first European settlement?

_____

14. Where is Timbuktu? (Watch for variant spellings.)

_____

15. Name and locate the largest lake in Africa.

_____

16. Name the longest river system in Africa.

_____

17. Name and locate the highest mountain in Africa.

_____

18. What is the name of Cape Town's mountain?

_____

19. Where is Victoria Falls?

_____

20. What is the primary basis of Egypt's appeal to tour-
    ists? _____

_____

21. What is the primary basis of Kenya's appeal to tourists? _____

_____

22. Name the great dam on the Nile River.

_____

23. Name the oil-rich country of West Africa.

_____

24. Name four archaeological sites around Luxor.

_____

_____

_____

25. What and where are the following:

a. Serengeti Plains _____

b. Durban _____

c. Stellenbosch _____

d. Tsavo and Amboseli _____

e. Garden Route _____

f. Mombasa _____

g. Blue Train _____

h. Abu Simbel _____

i. Kruger _____

j. Khan El Khalili _____

k. Groot Constantia _____

NOTES

CHAPTER
12

HAWAII
AND
OCEANIA

HAWAIIAN ISLANDS

The Hawaiian Islands contain seven main islands—Hawaii, Oahu, Maui, Molokai, Lanai, Kauai, and Nihau—and are considered one of the great tourist destinations. They combine what for most people are the ingredients of paradise—beautiful beaches, lush tropical foliage, soaring mountains, and a balmy climate—with all the amenities of modern living. Situated 2,400 miles southwest of Los Angeles, Hawaii is at the crossroads of the Pacific and is convenient to Australian, New Zealand, North American, and Japanese tourists.

Honolulu, the largest city on the islands, is the center of commerce and transportation as well as tourism. It is located on *Oahu,* usually considered the main island. Tourism in Honolulu is centered around *Waikiki,* probably the most famous beach in the world. There, under the watchful eye of the famous landmark, *Diamond Head,* an extinct volcano, thousands of vacationers enjoy the sun, sand, and surf that has made Hawaii such a vacation paradise.

While most tourists to Honolulu are beach-oriented, the city offers other attractions as well: historic *Pearl Harbor,* scene of the attack which brought the United States into World War II; colorful *Chinatown;* and the *Iolani Palace,* home of the former monarchy and the only royal palace in the United States. Outside the Honolulu area, visitors to Oahu head for the *Polynesian Cultural Center,* a living museum of South Pacific cultures especially well known for its dance troupe, and to the giant surf at *Makaha.*

Waikiki Beach, Honolulu, Hawaii. *Photo: Courtesy Hawaii Visitors Bureau*

All the other islands are known as the *Neighbor Islands.* The three favorite Neighbor Islands are Maui, Kauai, and the island of Hawaii itself. Many people consider *Maui* to have the most beautiful beaches of all the islands. The resorts on this island are clustered and centered in two areas, *Kaanapali* and *Waimea. Lahaina,* an old whaling town, now a national historic landmark, provides a glimpse of Hawaiian history and has many stores for interesting shopping.

Visitors to Maui often take day trips to the *Seven Falls of Hana* and to the *Haleakala Crater.* The Seven Falls of Hana are reached by a stunning but arduous drive along the lush northern coast of the island. The reward for the several hours drive along this twisting road is a series of beautiful pools connected by waterfalls as a river cascades into the sea.

The road to the *Haleakala Crater* is interesting as well for it winds from the tropical coast to the alpine air of 10,000 feet. As you can well imagine, the views from the road and the summit are magnificent. At the top, visitors can view the immense crater of the world's largest dormant volcano. There is a visitors' center where displays, illustrated lectures, and brochures, all provided by the National Park Service, make the excursion much more interesting. Many visitors enjoy walking in the area and watching the colors of the crater change with the day's light. The spectacle of sunrise at Haleakala is extra special and always draws hardy souls willing to get up early and brave the chilly morning mountain air.

Kauai, often called the *Garden Island,* is the lushest of all the Hawaiian Islands. The resorts on this island are centered around *Poipu Beach.* Kauai is especially well known for its scenic splendors. Chief among these is the *Waimea Canyon,* a magnificent gorge, rightfully called the *Grand Canyon of the Pacific.* The remote *Na Pali Coast* is a wilderness area of incomparable beauty where rugged mountains drop straight down to the sea. It is a favorite area for backpackers. Tourists can also cruise up the *Wailua River* to see the *Fern Grotto.*

The island of *Hawaii,* commonly called the *Big Island* because it is larger than all the others combined, is also the most varied of the islands. Here, visitors can enjoy the lush tropical side of the island in resorts around *Hilo* or the dry desert side of the island around *Kailua Kona.*

Visitors can snow-ski on *Mauna Kea,* hike to *Akala Falls* near Hilo, and swim from black sand beaches. They can visit numerous historic sites, the only coffee plantation in the United States (Kona coffee), or a macadamia nut farm and factory. And they can drive to the southernmost point of the United States.

The highlight of a visit to the Big Island, though, is *Volcanoes National Park.* There the crater of the active volcano, *Kilauea,* is immediately accessible. Visitors can drive within 200 feet of its summit or all the way around it on good roads. This experience allows people to come face-to-face with the vast forces which shape our earth.

AUSTRALIA

Although the Hawaiian Islands are a part of the United States, they are often associated with the other islands of the *South Pacific.* There are thousands of such islands, ranging from coral atolls to the massive island-continent Australia, as well as New Zealand, Fiji, and Tahiti.

All of these islands, collectively called *Oceania,* form a continent different from any of the others. Given its size, population, and wealth, Australia is the dominant country of Oceania. In fact, it is often referred to as a continent in its own right.

Australia's climate, casual way of life, British heritage, English language, and scenic attractions have made it a favorite destination for North American tourists.

Sydney is the gateway to Australia and the major city of Oceania. Set on a magnificent harbor, the city is one of the most attractive and exciting in the world. Tourists head for the *Circular Quay* in the center of the Manhattan-like downtown area for ferryboat rides across the harbor to lush suburbs and glamorous beaches such as *Manly Beach.* The city's best-known landmark, the striking *Sydney Opera House,* juts into the beautiful harbor.

With such a setting, Sydney is a city for views, especially from fifty-story *Australia Square,* the tallest building in the Southern Hemisphere. The panoramic view takes in two of Sydney's most colorful areas. The *Rocks,* adjacent to the Circular Quay, was the site of the city's earliest

Ayers Rock, Australia. *Photo: Courtesy Australian Tourist Commission*

settlement, but became one of the roughest districts in the Seven Seas. Today, though, it is a restored historic area of boutiques, galleries, restaurants, and pleasant homes. On the other hand, *King's Cross* has retained its risqué reputation for nighttime entertainment and bohemian life.

In contrast to Sydney's exuberance and vitality, her arch rival to the south, *Melbourne,* prides herself on being more reserved and dignified. Truly, Melbourne is a handsome city of beautiful parks, fine shopping and cultural centers, with a distinctly European air. Nearby, the *Healesville Sanctuary* displays much of the unique Australian wildlife: emus, koalas, wombats, kangaroos, opossums, and even a platypus.

Brisbane, north of Sydney, is the center of Australia's major beach resort area known as the *Gold Coast.* Farther north is the *Great Barrier Reef,* the longest coral reef in the world, where many resorts can be found.

The *Outback* stretches west of the coastal mountain range. There, vast sheep ranches gradually give way to the desert which forms the *Dead Center* of Australia. *Alice Springs,* a small but famous town right in the middle of the Dead Center, serves as a jumping off place for tourists visiting remote cattle stations and viewing the sights of the outback such as *Ayers Rock,* the largest monolith in the world.

Other points of interest are *Canberra,* the nation's capital, planned largely after Washington, D.C., and the *Barossa Valley,* the Germanic wine country near Adelaide. Australia's scenic and historic island-state, *Tasmania,* is best known for *Port Arthur,* the ruins of an old penal colony that once held 12,000 convicts. Lastly, many visitors enjoy crossing Australia from Sydney to Perth on the world-famous *Indian Pacific Train.*

NEW ZEALAND

Most tourists to Australia also visit New Zealand. As with Australia, the British background and casual life-style are appealing to North Americans, but the real attraction of New Zealand is its scenic splendor. So magnificent and so varied is the scenery of this island-country that it is described as a combination of Alaska, Hawaii, England, Ireland, Norway, Japan, and Switzerland. Moreover, the native people of New Zealand, the *Maori,* have a rich and distinctive South Pacific culture.

Auckland, at the northern end of the *North Island,* is the country's largest city and gateway. Tourism on the North Island centers around *Rotorua,* an interesting thermal and volcanic area. The *South Island* dazzles visitors with the *Southern Alps,* topped by *Mount Cook,* its highest peak. *Queenstown* on beautiful *Lake Wakatipu* is the center of this alpine year-round playground. In addition to skiing, visitors can literally jet down rivers on New Zealand's river jet boats, cruise the lake, play golf, go hiking, visit a remote sheep station, or take excursions to other beautiful sites such as *Milford Sound.* Other visitors enjoy a farm stay on either island, where they can share rural life with the farm family.

Much of the natural beauty of the South Island is protected by national parks. The largest and best known of these is *Fjordland* which includes the *Milford Sound Trek,* described as the most beautiful walk in the world.

Mount Cook and Mount Tasman From Lake Matheson, New Zealand.
Photo: Courtesy New Zealand Government Tourist Office

FIJI AND TAHITI

Two smaller island-nations with more exotic appeal are also prime attractions in Oceania: Fiji, with its British background, and Tahiti, with its French background. The appeal and the attractions of these two countries are similar to those of Hawaii—sun, sand, and surf in a South Pacific paradise.

The main resort area of Fiji is close-by *Nadi,* the international airport. The major city in Tahiti is *Papeete,* but resorts are spread throughout the islands. In fact, James Michener has described *Bora Bora* as the most beautiful island in the world.

At what better place could we complete our journey, reflect on where we've been, and dream of where we'll go?

Now that we have an idea of *where in the world,* we'll close with a brief study of *when in the world.*

Melbourne, Australia

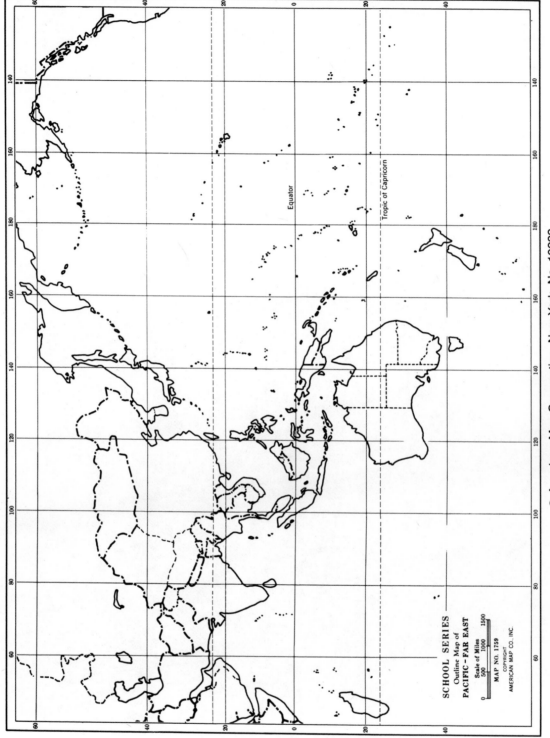

SCHOOL SERIES
Outline Map of
PACIFIC - FAR EAST

Scale of Miles

0   500   1000   1500

MAP NO. 1759
COPYRIGHT
AMERICAN MAP CO., INC.

Equator

Tropic of Capricorn

HAWAII AND OCEANIA

1. Name the main island of Hawaii. _____

2. Name the largest city of Hawaii. _____

3. What is the most famous beach in the world?

   _____

4. What is the only royal palace in the United States?

   _____

5. Name the three most popular Neighbor Islands.

   _____

   _____

6. Which is the largest Hawaiian island?

   _____

7. Which Hawaiian island is called the Garden Island?

   _____

8. On which island is the southernmost point of the United

   States? _____

9. Which Hawaiian island is said to have the most beauti-

   ful beaches? _____

10. On which Hawaiian island are the Seven Pools of Hana?

    _____

11. In which ocean is Oceania? _____

12. What is the major city of Oceania?

    _____

13. Name Australia's island-state.

    _____

14. What is the capital of Australia?

_____

15. Which city is the center of Australia's Gold Coast?

_____

16. Name the world-famous train which crosses Australia.

_____

17. What body of water separates Australia from New Zealand? _____

18. Name the two major islands of New Zealand.

_____

19. Which city is the gateway to New Zealand?

_____

20. What is the highest mountain in the Southern Alps?

_____

21. Name New Zealand's best-known park.

_____

22. Name two countries other than Australia and New Zealand in Oceania.

_____

23. What and where are the following:

a. Milford Sound Trek _____

b. Lahaina _____

    c. Sydney Opera House _____

    d. Kilauea _____

    e. Rotorua _____

    f. Ayres Rock _____

    g. Haleakala _____

    h. Diamond Head _____

    i. Great Barrier Reef _____

    j. King's Cross _____

    k. The Rocks _____

    l. Healesville Sanctuary _____

    m. Port Arthur _____

    n. Circular Quay _____

    o. Australia Square _____

24. Starting at Darwin, place these Australian cities in clockwise order: Adelaide, Brisbane, Canberra, Melbourne, Perth, Sydney.

    a. _____   e. _____

    b. _____   f. _____

    c. _____   g. _____

    d. _____

25. Name the city in the very center of Australia.

_____

NOTES

INTERNATIONAL TIME

When in the World?

Time is almost as important as *place* in the travel industry. Travel professionals work comfortably with the *24-Hour Clock* and the international time zone system called *Greenwich Mean Time* or *GMT*.

24-HOUR CLOCK

In North America, the 24-Hour Clock is used mainly by military and hospital personnel, but, in other areas of the world, it is used commonly by all. Because of its widespread use and greater precision, the North American travel industry uses it for international travel schedules and arrangements.

We commonly tell time by counting the hours around the clock twice and separating our two five o'clocks and our two ten o'clocks by a.m. and p.m. Users of the 24-Hour Clock only count the clock hours once. After reaching twelve noon, they continue on to thirteen, fourteen, and so on, instead of returning to one, two, etc. The 24-Hour Clock goes up to 22, 23, and 24 for 10 p.m., 11 p.m., and 12 a.m. (midnight). Instead of distinguishing hours by a.m. and p.m., they use a different number for each hour of the day. You can see that the 24-Hour Clock is more precise and has less chance for error. Below are the 24-Hour Clock equivalents to a.m./p.m. times.

A.M./P.M. CLOCK	24-HOUR CLOCK
12:00 a.m. (midnight)	0000 (or 2400) (midnight)
1:00 a.m.	0100
2:00 a.m.	0200
3:00 a.m.	0300
4:00 a.m.	0400
5:00 a.m.	0500
6:00 a.m.	0600
7:00 a.m.	0700
8:00 a.m.	0800
9:00 a.m.	0900
10:00 a.m.	1000
11:00 a.m.	1100
12:00 p.m. (noon)	1200 (noon)
1:00 p.m.	1300
2:00 p.m.	1400
3:00 p.m.	1500
4:00 p.m.	1600
5:00 p.m.	1700
6:00 p.m.	1800
7:00 p.m.	1900
8:00 p.m.	2000
9:00 p.m.	2100
10:00 p.m.	2200
11:00 p.m.	2300
12:00 a.m. (midnight)	2400 (or 0000) (midnight)

MINUTES AND THE 24-HOUR CLOCK

Minutes are always given following the hour as on a digital clock. Some examples: 0043, 1015, 1603, 2258.

Note that while you may write the stroke of midnight itself as 2400 *or* 0000, the minutes after midnight are always written with 00. In point of fact, you may not go beyond the 2400 in any given day. You have to start over. The first minute after midnight must be 0001: it may not be 2401. Our 12:15 a.m. must be 0015 and not 2415.

Lastly, 24-Hour Clock times are always written with four figures (as the various examples we have used illustrate). When given verbally, we also use all four figures, so that 0748 is stated as oh-seven-forty-eight, 1509 is fifteen-oh-nine, 2258 is twenty-two-fifty-eight, and 1900 is nineteen-hundred.

Complete the exercises below to make sure you understand the use of the 24-Hour Clock before you go on to Greenwich Mean Time.

1. Convert the following a.m./p.m. clock times to 24-Hour Clock times:

   a. 2:00 p.m. _____   g. 4:15 a.m. _____

   b. 4:15 p.m. _____   h. 1:05 a.m. _____

   c. 10:33 p.m. _____   i. 12:00 p.m. _____

   d. 7:20 a.m. _____   j. 12:00 a.m. _____

   e. 7:20 p.m. _____   k. 12:13 a.m. _____

   f. 11:41 p.m. _____   l. 12:42 p.m. _____

2. Convert the following 24-Hour Clock times to a.m./p.m. clock times:

   a. 0700 _____   g. 0200 _____

   b. 1900 _____   h. 0101 _____

   c. 1004 _____   i. 0001 _____

   d. 1537 _____   j. 1819 _____

   e. 1215 _____   k. 0734 _____

   f. 0030 _____   l. 2359 _____

GREENWICH MEAN TIME

By international agreement, the world has been divided into 24 time zones, each running from the North Pole to the South Pole. You can visualize this system by picturing a whole peeled orange which has 24 sections. Each one of those 24 sections is one time zone. Each of the 24 time zones is one hour from its neighboring time zones. At any given hour then, it is one of the 24 hours somewhere in the world.

All time zones have designations according to their relationship to the *prime meridian,* an imaginary line on the globe which runs through Greenwich, England, just outside London. Time at the prime meridian is known as Greenwich Mean Time or *GMT* and may be regarded as the "standard time" of the world since all other time zones are measured against it. The time zone through which the prime meridian actually passes is on Greenwich Mean Time and is the 0 (zero) time zone or simply GMT.

The twelve time zones to the east of GMT are ahead of it, and have a plus (+) designation, the twelve to the west are behind it and have a minus (−) designation. For instance, the time zone where Paris, France, (east of GMT) is located is designated as +1 GMT, meaning that the time there is one hour ahead of GMT. So when it is 1200 (noon) in GMT, it is 1300 in Paris. Athens, Greece is farther east yet and is in +2 GMT, meaning that the time there is two hours ahead of GMT. So when it is 1200 in GMT and 1300 in Paris (+1 GMT), it is 1400 in Athens. We can continue traveling east through Pakistan (+5 GMT), Japan (+9 GMT), to New Zealand (+12 GMT). When it is 1200 GMT, 1300 in Paris (+1 GMT), 1400 in Athens (+2 GMT), it is 1700 in Pakistan (+5 GMT), 2100 in Japan (+9 GMT), and 2400, or midnight, in New Zealand (+12 GMT).

The same system operates going west of GMT. For instance, New York City is in the −5 GMT zone, meaning that the time there is five hours behind GMT. So, when it is 1200 in GMT, it is 0700 in New York City (−5 GMT). We can continue west through Los Angeles (−8 GMT or eight hours behind GMT) to Honolulu (−10 GMT or ten hours behind GMT) all the way to the −12 GMT zone in the middle of the Pacific Ocean. So, when it is 1200 in GMT and 0700 in New York City (−5 GMT), it will be 0400 in Los Angeles (eight hours behind GMT), 0200 in Honolulu, ten hours behind GMT, and midnight in the mid-Pacific (twelve hours behind GMT).

You may have counted GMT (0 time zone), added twelve plus time zones and twelve minus time zones and come up with a total of 25 instead of the 24 we introduced. Keep in mind that −12 GMT and +12 GMT are actually in one zone, split by the *International Date Line.* The International Date Line is an imaginary line, directly opposite the globe from the prime meridian. The International Date Line divides one day (date) from another. Since the time zones on either side of it, +12 GMT and −12 GMT, are always 24 hours apart, they will always be on different days (dates).

GMT CHART

To help make this system clearer to you, we have included a *GMT Chart* on page 168. Turn to it now. Note that each column is designated as one of the GMT zones at the top. Note also that we have listed some of the areas of the world which are in particular time zones.

To follow the example we used above, find 1200 at the bottom of the GMT or 0 column. That represents 1200 at GMT. Then move your finger horizontally on the same level as the 1200 in the GMT column to the +1 GMT column. There you will see that it reads 1300, telling you that when it is 1200 in GMT, it is 1300 in +1 GMT. Now move your finger horizontally to the +2 column where it will read 1400 (or 2 hours ahead of GMT), so that when it is 1200 in GMT, it is 1400 in +2 GMT.

After you have finished tracing the day across the plus zones to +12 GMT, New Zealand, return to GMT and trace the day horizontally across the minus zones through New York (−5 GMT), Los Angeles (−8 GMT), Honolulu (−10 GMT), and the mid-Pacific (−12 GMT).

Now ask yourself how the 0000 of the +12 GMT column differs from the 0000 of the −12 GMT column. The answer is that in the midnight at +12 GMT, the day we traced is just ending and the next day is beginning, whereas in the midnight of the −12 GMT, the day we traced is just beginning and the day before is just ending.

You will see that each horizontal line on the GMT Chart has all the hours of a single day. Once you know what time it is in any one zone of the world, you can use this chart to determine the time in any other time zone. For instance, let us say that we want to know the time in Honolulu when it is 1700 in New York. First find the −5 GMT column where New York is located, trace your finger horizontally on the same horizontal level as the 1700 in the −5 GMT column until you come to the −10 GMT column and you will find 1200. So, when it is 1700 in New York, it is 1200 in Honolulu. Complete the exercise on page 169 to practice using the GMT Chart.

TIME ZONES IN THE CONTIGUOUS UNITED STATES

We may now approach the time zones which cross the contiguous United States with more understanding. Most of you know the names of those four standard time zones: *Eastern* or *EST* for Eastern Standard Time, *Central* or *CST*, *Mountain* or *MST*, *Pacific* or *PST*. Calling on our knowledge of GMT, let us add the proper designations to each of these zones, EST is −5 GMT, CST is −6 GMT, MST is −7 GMT, and PST is −8 GMT.

Logically enough, if EST is five hours behind GMT and PST is eight hours behind GMT, PST must be three hours behind EST as is the case with Los Angeles (in PST) and New York (in EST).

Complete the exercise below to underscore your understanding of the relationship of these time zones to one another:

1. EST is how many hours ahead of CST _____, MST

_____, PST _____?

2. CST is how many hours behind EST _____, ahead of

MST _____, and ahead of PST _____?

3. MST is how many hours behind EST _____ , CST

_____ , and how many hours ahead of PST _____ ?

4. PST is how many hours behind EST _____ , CST

_____ , and MST _____ ?

ELAPSED FLYING TIME

Our knowledge of GMT will help us to understand international flight schedules and the actual amount of flying time, called the elapsed flying time, of an air journey. This information helps us better serve our clients, since we can suggest the most appropriate flight schedules for their business or vacation plans.

To determine elapsed flying time, we simply subtract the departure time of a flight from the arrival time of a flight. The answer gives us the amount of time that has elapsed between departure and arrival—the elapsed flying time. For instance, let us suppose that your flight departs New York at 0800 (or 8:00 a.m.) and arrives in Miami at 1100 (or 11:00 a.m.). Since both New York and Miami are in the *same* time zone (−5 GMT), it is an easy matter to subtract 0800 from 1100 and determine three hours of elapsed flying time.

When the departure city and the arrival city are in *different* time zones, however, we must make some adjustment before subtracting, because subtracting one time zone from another is like subtracting apples from oranges. There are a number of ways to make the adjustment so that we are subtracting apples from apples or oranges from oranges, that is, so that we are subtracting one time from another in the *same* time zone. We find that the simplest method is to convert both the departure time and the arrival time to their GMT equivalents and then subtract the converted departure time from the converted arrival time.

For instance, let us say that your client's flight departs New York at 0800 and arrives in Paris at 2100. Simply subtracting 0800 from 2100 would give us 13 hours of elapsed flying time, a most incorrect answer (based on subtracting time in the −5 GMT from a time in the +1 GMT, or apples from oranges). To rectify that situation and to arrive at the correct answer, convert both the departure time and the arrival time to their equivalent times in GMT (0 time zone) and then subtract. In other words, determine what time it is in GMT when you depart New York. Then determine what time it is in GMT when you arrive in Paris. Those times in GMT are your converted departure and arrival times. You may now subtract the converted departure time from the converted arrival time because both are in the *same* time zone, GMT (or 0 time zone).

By using the GMT chart as you did in an earlier exercise, you can determine that when it is 0800 in New York (−5 GMT), it is 1300 in GMT, so we can call 1300 the GMT equivalent to the departure time. Likewise, by using the GMT chart, we can determine that when it is 2100 in Paris (+1 GMT), it is 2000 in GMT, so we can call 2000 the GMT equivalent to the arrival time. We have now converted both departure and arrival times to GMT. Since both our departure time and our arrival times are now in the same time zone, we may subtract. Since GMT is, in a sense, the standard time of the world, and since all time zones are designated according to their relationship to GMT, what better time to use for this purpose?

REVIEW

Arrival time in Paris: 2100,   converted to GMT equivalent: 2000

minus

Departure time in New York: 0800,   converted to GMT equivalent: <u>1300</u>

The elapsed flying time is **7 hours.**   0700

This procedure will work for all flights that cross time zones regardless of direction. Just follow these rules.

1. Always work in 24-Hour Clock time. If the departure and arrival times are given to you in a.m. or p.m., convert them to their 24-Hour Clock equivalents.
2. Use your GMT Chart to convert both the departure time and the arrival time to GMT equivalents.
3. Subtract the converted departure time from the converted arrival time.
4. You may sometimes wind up with a problem like this one:

     1620
     −1040

   Remember that you have to subtract the 40 from the 20 because they are minutes and the 10 from the 16 because they are hours. How to subtract 40 minutes from 20 minutes? Borrow 60 minutes from the 16 hours, make the 16 hours 15 hours and the 20 minutes 80 minutes. Then the problem above will look like this:

     1580
     −1040

   540 or 5 hours and 40 minutes of elapsed flying time.
5. Many international flights land the day after they depart. For instance, Olympic Airways flight 412 departs New York at 1845 and arrives in Athens at 1050 + 1 (which means at 1050 the day after the departure day). To determine the elapsed flying time in this case, proceed as before—convert the departure and the arrival times to their GMT equivalents. Since Athens is +2 GMT, 1050 +1 becomes 0850 + 1. Our math problem then becomes:

     0850 + 1 (converted arrival time +1)
     −2300   (converted departure time)

   Our next step is to recognize that the +1 means a day later so we convert that to 24 hours and add the 24 hours to the 0850. Our math problem then becomes:

     3250 (0850 + 2400)
     −2300

   950 or 9 hours and 50 minutes of elapsed flying time.

EXAMPLE 1

Depart from Rome 1200
Arrive in Chicago 1645

1. Convert the departure and arrival times to their GMT equivalents. Rome is +1 GMT and Chicago is −6 GMT, so the converted Rome departure time is 1100, the converted Chicago arrival time is 2245.
2. Subtract the converted departure time from the converted arrival time:

  2245
  −1100
  ────
  1145 or 11 hours and 45 minutes of elapsed flying time.

EXAMPLE 2

Depart from Los Angeles 1130
Arrive in Bangkok, Thailand 2230 + 1

1. Convert the departure time and arrival times to their GMT equivalents. Los Angeles is −8 GMT, Bangkok is +7 GMT, so the converted departure time is 1930 and the converted arrival time is 1530 +1.
2. Convert the +1 to 24 hours, add the 24 hours to the 1530, for a total of 3930.
3. Subtract the departure time from the converted arrival time:

  3930
  −1930
  ────
  2000 or 20 hours of elapsed flying time.

INTERNATIONAL DATE LINE

GMT CHART

July 4   July 5

GMT	Location																								
−12	Mid Pacific	2300	2200	2100	2000	1900	1800	1700	1600	1500	1400	1300	1200	1100	1000	0900	0800	0700	0600	0500	0400	0300	0200	0100	**0000**
−11		**0000**	2300	2200	2100	2000	1900	1800	1700	1600	1500	1400	1300	1200	1100	1000	0900	0800	0700	0600	0500	0400	0300	0200	0100
−10	Hawaii	0100	**0000**	2300	2200	2100	2000	1900	1800	1700	1600	1500	1400	1300	1200	1100	1000	0900	0800	0700	0600	0500	0400	0300	0200
−9		0200	0100	**0000**	2300	2200	2100	2000	1900	1800	1700	1600	1500	1400	1300	1200	1100	1000	0900	0800	0700	0600	0500	0400	0300
−8	L.A./S.F./PST	0300	0200	0100	**0000**	2300	2200	2100	2000	1900	1800	1700	1600	1500	1400	1300	1200	1100	1000	0900	0800	0700	0600	0500	0400
−7	Denver/PST	0400	0300	0200	0100	**0000**	2300	2200	2100	2000	1900	1800	1700	1600	1500	1400	1300	1200	1100	1000	0900	0800	0700	0600	0500
−6	Chicago/CST	0500	0400	0300	0200	0100	**0000**	2300	2200	2100	2000	1900	1800	1700	1600	1500	1400	1300	1200	1100	1000	0900	0800	0700	0600
−5	NYC/EST	0600	0500	0400	0300	0200	0100	**0000**	2300	2200	2100	2000	1900	1800	1700	1600	1500	1400	1300	1200	1100	1000	0900	0800	0700
−4	Bermuda	0700	0600	0500	0400	0300	0200	0100	**0000**	2300	2200	2100	2000	1900	1800	1700	1600	1500	1400	1300	1200	1100	1000	0900	0800
−3		0800	0700	0600	0500	0400	0300	0200	0100	**0000**	2300	2200	2100	2000	1900	1800	1700	1600	1500	1400	1300	1200	1100	1000	0900
−2		0900	0800	0700	0600	0500	0400	0300	0200	0100	**0000**	2300	2200	2100	2000	1900	1800	1700	1600	1500	1400	1300	1200	1100	1000
−1		1000	0900	0800	0700	0600	0500	0400	0300	0200	0100	**0000**	2300	2200	2100	2000	1900	1800	1700	1600	1500	1400	1300	1200	1100
0	U.K./Ireland/Portugal	1100	1000	0900	0800	0700	0600	0500	0400	0300	0200	0100	**0000**	2300	2200	2100	2000	1900	1800	1700	1600	1500	1400	1300	1200
+1	Most of Western Europe	1200	1100	1000	0900	0800	0700	0600	0500	0400	0300	0200	0100	**0000**	2300	2200	2100	2000	1900	1800	1700	1600	1500	1400	1300
+2	Israel/Egypt/Greece	1300	1200	1100	1000	0900	0800	0700	0600	0500	0400	0300	0200	0100	**0000**	2300	2200	2100	2000	1900	1800	1700	1600	1500	1400
+3	Saudi Arabia	1400	1300	1200	1100	1000	0900	0800	0700	0600	0500	0400	0300	0200	0100	**0000**	2300	2200	2100	2000	1900	1800	1700	1600	1500
+4		1500	1400	1300	1200	1100	1000	0900	0800	0700	0600	0500	0400	0300	0200	0100	**0000**	2300	2200	2100	2000	1900	1800	1700	1600
+5	Pakistan	1600	1500	1400	1300	1200	1100	1000	0900	0800	0700	0600	0500	0400	0300	0200	0100	**0000**	2300	2200	2100	2000	1900	1800	1700
+6		1700	1600	1500	1400	1300	1200	1100	1000	0900	0800	0700	0600	0500	0400	0300	0200	0100	**0000**	2300	2200	2100	2000	1900	1800
+7	Thailand	1800	1700	1600	1500	1400	1300	1200	1100	1000	0900	0800	0700	0600	0500	0400	0300	0200	0100	**0000**	2300	2200	2100	2000	1900
+8	Hong Kong	1900	1800	1700	1600	1500	1400	1300	1200	1100	1000	0900	0800	0700	0600	0500	0400	0300	0200	0100	**0000**	2300	2200	2100	2000
+9	Japan	2000	1900	1800	1700	1600	1500	1400	1300	1200	1100	1000	0900	0800	0700	0600	0500	0400	0300	0200	0100	**0000**	2300	2200	2100
+10	Eastern Australia	2100	2000	1900	1800	1700	1600	1500	1400	1300	1200	1100	1000	0900	0800	0700	0600	0500	0400	0300	0200	0100	**0000**	2300	2200
+11		2200	2100	2000	1900	1800	1700	1600	1500	1400	1300	1200	1100	1000	0900	0800	0700	0600	0500	0400	0300	0200	0100	**0000**	2300
+12	New Zealand	2300	2200	2100	2000	1900	1800	1700	1600	1500	1400	1300	1200	1100	1000	0900	0800	0700	0600	0500	0400	0300	0200	0100	**0000**

INTERNATIONAL TIME

When in the World?

USING THE GMT CHART

1. What time is it in Los Angeles when it is 1800 in New York? _____

2. What time is it in Tokyo, Japan when it is 0200 in Athens, Greece? _____

3. What time is it in Paris, France when it is 1100 in Chicago? _____

4. What time is it in Denver when it is 2100 in Cairo, Egypt? _____

5. What time is it in Karachi, Pakistan when it is 1600 in London, U.K.? _____

6. What time is it in San Francisco when it is 1600 in London, U.K.? _____

7. What is the day and date in Sydney, Australia when it is 1000, July 4, in Auckland, New Zealand?

    _____

8. What is the day and date in Honolulu when it is 2200, July 4, in Bermuda? _____

9. What is the day and date in New York when it is 0700, July 5, in Hong Kong? _____

10. What is the day and date in Bangkok, Thailand when it is 1900 July 4, in Chicago?

_____

11. What time and date is it in Tel Aviv, Israel when it is 1300, September 9, in Melbourne, Australia?

_____

12. What time and date is it in Frankfurt, West Germany when it is 1900, April 5, in New York?

_____

CONVERTING TIME TO GMT

Practice converting time to GMT in this exercise before tackling the Elapsed Flying Time Exercise which follows. Use your GMT chart and ask yourself:

1. What time is it in GMT when it is 2300 in Paris, France?

_____

2. What time is it in GMT when it is 1400 in Tokyo, Japan? _____

3. What time is it in GMT when it is 0500 in Honolulu?

_____

4. What time is it in GMT when it is 1100 in New York?

_____

5. What time is it in GMT when it is 1700 in Auckland, New Zealand? _____

6. What time is it in GMT when it is 2030 in Hong Kong?

_____

7. What time is it in GMT when it is 0315 in Tel Aviv, Israel? _____

8. What time is it in GMT when it is 1643 in Bermuda?

   _____

9. What time is it in GMT when it is 1358 in Dublin, Ireland? _____

10. What time is it in GMT when it is 2210 in Chicago?

   _____

ELAPSED FLYING TIME PROBLEMS

Determine the elapsed flying times for each of the following flights.

1. Delta 842
   Departs from San Francisco at 0705
   Arrives in Los Angeles at 0815   _____

2. Pan American 130
   Departs from New York at 1000
   Arrives in Bermuda at 1255   _____

3. Continental 3
   Departs from Denver at 0535
   Arrives in Honolulu at 1048   _____

4. KLM 642
   Departs from New York at 1800
   Arrives in Amsterdam, Netherlands at
   0705 + 1   _____

5. Lufthansa 430
   Departs from Frankfurt, West Germany at 1300
   Arrives in Chicago at 1520   _____

6. Pan American 5
   Departs from San Francisco at 1415
   Arrives in Hong Kong at 2055 + 1   _____

7. Japan Airlines 066
   Departs from Tokyo, Japan at 1340
   Arrives in Los Angeles at 0600   _____

8. British Airways 32
   Departs from London, U.K. at 1155
   Arrives in Lisbon, Portugal at 1430   _____

NOTES

GLOSSARY

ABC'S of South America. Code for the divisions of South America. **A** for Andes countries, **B** for Brazil, **C** for Caribbean countries, **S** for Southern countries. (*See* chapter 5, South America, for fuller explanation.)

Aconcagua. Highest mountain in the Andes, 22,834 feet.; located in Argentina near the Chilean border.

Aegean Sea. Branch of the Mediterranean Sea, between Greece and Turkey; surrounds most of the Greek Islands.

Africa. *See* **Continents.**

Alps. Major mountain range of Western Europe, centered in Switzerland, extending into France, Italy, and Austria.

Amazon River. Major river system of South America, largest river in world by volume; drains world's greatest rain forest.

Andes. Major mountain range of South America extending along the entire length of the west coast.

Anglo-America. Those areas of North America with a British rather than a Latin heritage; the United States and Canada (except Quebec).

Antarctica. *See* **Continents.**

Appalachian. Major mountain range of Eastern North America extending from the Canadian Atlantic Provinces to Alabama; includes Green Mountains and White Mountains in New England, Catskills in New York, Alleghenies in Pennsylvania, Virginia, Maryland, and West Virginia.

Arabian Peninsula. Southwest Asia, bounded by the Red Sea, Arabian Sea, and Persian Gulf, largely occupied by Saudi Arabia; a major oil-producing region.

Archipelago. A group of islands.

Arctic Ocean. The ocean north of the Arctic Circle. *See also* **Ocean.**

Asia. *See* **Continents.**

Atlantic Ocean. The ocean between North and South America and Europe and Africa. *See also* **Ocean.**

Atlantic Provinces. The four eastern Canadian provinces: New Brunswick, Newfoundland, Nova Scotia, and Prince Edward Island. *See also* **Maritime Provinces.**

Australia. Sometimes considered a continent in its own right, it is also combined with the other islands of the South Pacific, including New Zealand; all the islands together are called Oceania. *See also* **Continents.**

Ayers Rock. Largest monolith in the world, central Australia.

Baja California. Also called Lower California, peninsula in northwest Mexico, extending from California, bounded by Pacific Ocean and Gulf of California.

Balkans. The countries occupying the Balkan Peninsula in southeast Europe, bounded by the Adriatic, Ionian, Mediterranean, Aegean, and Black Seas; includes Albania, Bulgaria, Greece, Romania, Yugoslavia, and European Turkey.

Bay or **Gulf.** A body of water partially enclosed by land that has an opening to a lake, sea, or ocean.

Benelux. Combined form of Belgium, Netherlands, and Luxembourg.

Bering Strait. Passage of water separating Asia and North America, fifty-three miles wide at its narrowest point.

Bight. A curve in a coastline similar to a bay.

Blanc, Mont. Highest mountain in the Alps and in Western Europe, 15,781 ft.; located in France, near the border with Italy and Switzerland.

Bosporus. Straits connecting the Black Sea with the Sea of Marmara and ultimately the Mediterranean Sea; separates European Turkey from Asian Turkey.

British Isles. Major island group off the coast of Western Europe; includes both Great Britain and Ireland.

Canadian Shield. Large, rugged area forming a horseshoe around Hudson Bay, Canada. Barren in the north, forested in the south; only lightly settled, divides settled Canada into eastern and western areas.

Canyon. A narrow gorge eroded into the earth's surface by running water.

Cape. A piece of land which projects into a body of water.

Caribbean Sea. Branch of the Atlantic Ocean, surrounded by the West Indies, Central America, and the northeast coast of South America.

Central America. The area of North America south of Mexico; includes Guatemala, Honduras, El Salvador, Nicaragua, Costa Rica, and Panama.

Central Corridor of Europe. The general route followed by most tour companies on their grand tours of Europe; includes London, Paris, the Swiss Alps, Florence, Rome, Venice, the Austrian Alps, Munich, the Rhine River Valley, Amsterdam. This tour route minimizes the distance covered but maximizes the cultures visited.

Coast. Land beside or near an ocean or sea.

Continental Divide. The ridge of the Rocky Mountains that separates rivers flowing east to the Atlantic Ocean and the Gulf of Mexico from those flowing west toward the Pacific.

Continents. The large landmasses of the earth including North and South America, Europe, Asia, Africa, Oceania, and Antarctica.

Cook, Mount. The highest mountain in Oceania, 12,349 feet, in the Southern Alps of New Zealand.

Dead Center. Arid area of Central Australia.

Dead Sea. A salty lake between Israel and Jordan, its surface is 1,302 feet below the Mediterranean Sea, making it the lowest point on the earth's surface.

Death Valley. Located in eastern California and southern Nevada; contains Badwater, a small pool 282 feet below sea level, the lowest point in the United States.

Delta. Deposits of soil formed at the mouth of a river.

Desert. A dry and barren area.

East. (1) Northeast region of the United States; includes New England and the Mid-Atlantic States, sometimes called the Northeast. (2) In world usage it refers to Asia. *See also* **Far, Middle** and **Near East** and **Orient.**

East Africa. Area of Africa centered in Kenya and Tanzania.

East Coast. Area of the United States along the Atlantic coast from Maine to District of Columbia; generally includes the chain of cities from Boston to Washington. *See also* **Northeastern Corridor.**

Eastern Europe. Area east of Scandinavia, West Germany, and Austria; includes the Balkans, except Greece. Once a division based on languages, history, and cultural traditions, now largely based on politics.

Eastern Hemisphere. The half of the earth which contains Europe, Asia, Africa, and Oceania.

English Channel. Sometimes called The Channel, a strait of water separating England in the British Isles from France on the mainland of Europe.

Equator. Imaginary line around the earth at its greatest width, 24,900 miles, equally distant from the North and South Poles; divides the earth into Northern and Southern Hemispheres.

Estuary. The wide mouth of a river where the current meets the tide.

Europe. *See* **Continents.**

Everest, Mount. Highest mountain in Asia and the world, 29,028 feet; located in the Himalayas, Nepal.

Far East. Area of Eastern Asia including Mongolia, China, Taiwan, Hong Kong, North and South Korea, and Japan. The nations of Southeast Asia are sometimes included by some geographers. *See also* **East** and **Southeast Asia.**

Fjord. A narrow inlet of an ocean or sea, often bordered by steep cliffs.

Fuji, Mount. Highest mountain in Japan, 12,388 feet; symbol of the country.

Ganges. Sacred river of Hindus in northeast India.

Gap. *See* **Pass.**

Gibraltar, Strait of. Passage of water connecting the Atlantic Ocean and the Mediterranean Sea, separating Europe and Africa; one of the world's most crucial points.

Glacier. A huge mass of permanent ice and snow.

Good Hope, Cape of. Peninsula on the southwest coast of South Africa, traditionally, but falsely considered the southernmost point of Africa.

Grand Canyon. Vast gorge of the Colorado River in northern Arizona, more than a mile deep in some places.

Great Barrier Reef. Coral reef off the northeast coast of Australia, largest deposit of coral in the world; largest living thing in the world.

Great Basin. High, dry plateau east of the Sierra Nevada mountains, United States; centered in Nevada, it extends into parts of California, Oregon, Idaho, Wyoming, and Utah.

Great Lakes. Chain of vast lakes in central North America comprised of Lakes Erie, Huron, Michigan, Ontario, and Superior. All but Michigan, which is wholly within the United States, are shared by Canada and the United States.

Great Lakes States. States bordering the Great Lakes, usually included in the Midwest Region, except for New York and Pennsylvania, which are usually included in the East.

Great Plains. The plains of the United States between the Mississippi Valley and the Rocky Mountains.

Great Triad of European Cities. London, Paris, and Rome.

Gulf. *See* **Bay.**

Himalayas. World's highest mountain range, centered in Nepal, Asia; separates the Indian Subcontinent from much of the rest of Asia.

Horn, Cape. The southernmost point of South America.

Hub of Europe. Switzerland because (1) it is the meeting place of three of Europe's major cultures, French, Italian, and German, and (2) it is centrally located in Western Europe.

Iberia. Region of southwest Europe occupying the Iberian Peninsula, bounded by the Mediterranean Sea, Atlantic Ocean, and Bay of Biscay and separated from the rest of Europe by the Pyrenees Mountains; includes Spain, Portugal, Andorra, and Gibraltar.

Iguazú Falls. Spectacular falls, two and one-half miles wide; located at the Brazilian-Argentine-Paraguayan border in South America.

Indian Subcontinent. Region of Asia, east of the Middle East and west of Southeast Asia and the Far East; separated from much of the rest of Asia by the Himalayas across its north; includes India, Pakistan, Bangladesh, Sri Lanka, Nepal, and several smaller Himalayan nations.

Indochina. Area of Southeast Asia occupying Indochinese Peninsula; includes Burma, Laos, Malaysia, Thailand, Vietnam, Singapore, and Kampuchea (formerly Cambodia).

Inside Passage. Cruise route from Seattle, Washington, and Vancouver, British Columbia, to Skagway, Alaska; located between the mainland of North America and its many offshore islands.

International Date Line. An imaginary line drawn from the North Pole to the South Pole through the Pacific Ocean at which each calendar day begins at midnight.

Island. A landmass, smaller than a continent, surrounded by water.

Isthmus. A narrow strip of land connecting two larger masses of land.

Kilimanjaro, Mount. Highest mountain in Africa, 19,340 feet; located in Tanzania near border with Kenya.

Korean Peninsula. East Asia, occupied by North and South Korea.

Lake. A body of fresh water enclosed by land.

Latin America. Located in the Western Hemisphere south of the Rio Grande River; includes Mexico, Central America, West Indies, and South America where cultures stem from a Latin (French, Spanish, or Portuguese) heritage.

Latitude. Any of the imaginary east-west lines that circle the globe providing, along with longitude, a means of locating any point on the earth's surface. *See also* **Longitude.**

Longitude. Any of the imaginary lines which run from pole to pole across the surface of the globe providing, along with latitude, a means of locating any point on the earth's surface. *See also* **Latitude, Meridian,** and **Prime Meridian.**

Magellan, Straits of. A water passage through the southern tip of South America that connects the South Atlantic and South Pacific Oceans.

Maritime Provinces. Three eastern provinces of Canada: New Brunswick, Nova Scotia, Prince Edward Island, but not Newfoundland. *See also* **Atlantic Provinces.**

McKinley, Mount. Highest mountain in the United States and North America, 20,320 feet; located in south central Alaska.

Mediterranean Sea. The body of water almost entirely surrounded by Europe, Asia, and Africa. It contains many branches including the Aegean and Adriatic Seas and is connected to the Atlantic Ocean by the Strait of Gibraltar.

Mediterranean World. The countries surrounding the Mediterranean Sea in Europe, Asia, and Africa.

Meridian. Any of the lines of longitude.

Mexican Riviera. The west coast of the "mainland" of Mexico where many resort cities are located.

Mid or Middle Atlantic States. Subdivision of the East (United States); includes New York, New Jersey, Pennsylvania, Maryland, Delaware, and District of Columbia. *See also* **East.**

Mideast or Middle East. Southwestern region of Asia from the Mediterranean Sea to and including Afghanistan as well as the Arabian Peninsula. At one time, the Mideast and the Near East were separate regions, but contemporary politics has led to a more inclusive use of the term Mid- or Middle East.

Mid or Middle West. North central region of the United States, north of the Ohio River. This area ranges from Ohio in the east to Kansas, Nebraska, and the Dakotas in the west; also includes the Great Lakes States.

Mississippi-Missouri-Ohio River System. The major river system of the United States; one of the great river systems of the world. It runs through the central United States between the Rockies and the Appalachians.

Mountain. A natural elevation, of great height, of the earth's surface.

Mountain West. Subdivision of the West (United States), the Rocky Mountain States; includes Nevada, Colorado, Utah, Wyoming, Montana, Idaho; usually regarded as the last of the "Old West."

Near East. Region of Southwest Asia, including the countries on the Arabian Peninsula that border the Mediterranean, Jordan and Syria. Generally combined with Iraq, Iran, and Afghanistan and termed the Mideast. *See also* **Mideast.**

New England. Subdivision of the East (United States), six states in the extreme northeast corner of the United States; includes Maine, Massachusetts, New Hampshire, Vermont, Rhode Island, and Connecticut.

Nile River. Located in Northeast Africa, the longest river in the world; closely associated with Egypt and its ancient civilization.

North Africa. The region of Africa which borders the Mediterranean Sea, separated from the rest of Africa by the Sahara Desert; includes Egypt, Tunis, Libya, Algeria, and Morocco.

North America. *See* **Continents.**

North Cape. The northernmost part of Europe, Northern Norway.

Northeast. *See* **East** (1).

Northeastern Corridor. The chain of cities in northeastern United States from Washington, D.C. to Boston, Massachusetts.

Northern Hemisphere. The half of the earth north of or above the equator. *See also* **Equator.**

Ocean. One of the four large bodies of salt water which cover the earth's surface; the Pacific, Atlantic, Indian, and Arctic Oceans with numerous bays, gulfs, and seas.

Oceania. *See* **Continents.**

Orient. Region of Asia that includes the Far East and Southeast Asia.

Outback. The lightly-settled, dry, flat interior of Australia.

Pacific Northwest. Subdivision of the West (United States); northwest corner, centered in Washington and Oregon, may include Alaska and very northern section of California.

Pampas. The great grasslands of South America, particularly in Argentina.

Panama, Isthmus of. The narrow bridge of land connecting North America with South America; the southernmost part of North America.

Panama Canal. Connects the Pacific Ocean with the Caribbean Sea through the Isthmus of Panama.

Pass. A passage or opening through mountains.

Peninsula. A long projection of land nearly surrounded by a body of water and connected to a larger landmass.

Plain. A level, treeless expanse of land.

Plateau. A high, level expanse of land.

Pole. Either end of the earth's axis.

Prairie. A level, or gently-rolling, expanse of grassland.

Prairie Provinces. Canadian provinces of Alberta, Saskatchewan, and Manitoba, occupying the southcentral region of the country between the Rockies and the Great Lakes.

Prime Meridian. The meridian which passes through Greenwich, England from which all points of longitude are measured.

Promontory. *See* **Cape.**

Rhine River. Prominent river that flows through central Europe from Switzerland to the North Sea.

Rio Grande. The river that separates Mexico and Texas, United States.

River. A natural stream of water which empties into another body of water.

Riviera.   The Mediterranean coast in southeast France and northeast Italy, noted for its scenery and climate.

Rocky Mountains.   Major mountain range in western North America, extending from New Mexico to Alaska. Often called the Rockies.

Sahara.   The desert that stretches across Northern Africa, ranging from the Atlantic coast to the Red Sea; the world's most extensive desert.

Savanna or **Savannah.**   A plain in a tropical or subtropical region. *See also* **Plain.**

Saint Lawrence.   The major river of eastern Canada; flows from Lake Ontario to the Gulf of the St. Lawrence.

Scandinavia.   The region of northwestern Europe that includes Denmark, Sweden, Norway and usually Finland and Iceland.

Sea.   Body of salt water, smaller than the oceans; often extensively enclosed by land.

Serengeti.   The vast plain in Tanzania, Africa, noted for its wildlife.

Sound.   A wide strait that links two large bodies of water or separates an island from the mainland.

South.   Southeast and South Central region of the United States from Virginia in the east to Oklahoma and Texas in the west, south of the Ohio River and the Missouri-Arkansas border.

South America.   *See* **Continents.**

Southeast Asia.   The region of Asia between southern China and the extreme eastern section of India including the nations of the Indochina Peninsula, Indonesia, and the Philippines.

Southern Alps.   The mountain range of South Island, New Zealand.

Southern Hemisphere.   The half of the earth south of or below the equator. *See also* **Equator.**

South Pacific.   The region of the Pacific Ocean which includes the islands of Oceania. *See also* **Oceania.**

Southwest.   The subdivision of the West (United States) that includes Arizona and New Mexico; sometimes includes west Texas.

Steppes.   The plains of the Soviet Union and Central Asia.

Strait.   A narrow passage of water which connects two other bodies of water and/or separates two points of land.

Suez Canal.   One of the world's most important waterways; connects the Red Sea with the Mediterranean Sea across the Isthmus of Suez in northeast Africa.

Tableland of Tibet.   A vast plateau in central Asia averaging 16,000 feet; the highest region of the world, often called the Roof of the World.

Titicaca, Lake.   At 12,500 feet, the highest navigable lake in the world, located on the border of Peru and Bolivia.

Ural Mountains.   The border between Europe and Asia, located in the Soviet Union.

Valley.   Low land between mountains or hills often with a river flowing through it.

Victoria Falls.   Spectacular falls of the Zambezi River on the border of Zambia and Zimbabwe, Africa.

West. (1) The western region of the United States including New Mexico, Colorado, Wyoming, Montana, and all the states located to the west of them. The largest of the four United States regions, subdivided into *Mountain West, Southwest, West Coast,* and *Pacific Northwest*; (2) Western Europe and the developed nations which grew from Western European countries, such as the United States.

Western Europe. Scandinavia, Greece, and the mainland of Europe west of East Germany, Czechoslovakia, and Hungary. Once a division based on languages, history, and cultural traditions, now largely based on politics.

Western Hemisphere. The half of the earth which contains North and South America.

West Coast. Subdivision of the West (United States) along the Pacific Coast from the Mexican border to the northern San Francisco Bay counties, California.

West Indies. The islands between North and South America separating the Atlantic Ocean from the Caribbean Sea.

Yangtze. The major river of China and one of the longest rivers in the world.

Yucatán. Peninsula in southeast Mexico that separates the Gulf of Mexico from the Caribbean Sea.

INDEX